SURVIVAL SENSE
FOR PILOTS AND PASSENGERS

by
Robert Stoffel and Patrick LaValla

EMERGENCY RESPONSE INSTITUTE

SPECIALISTS IN EMERGENCY PREPAREDNESS EDUCATION

Copyright 1980 by Emergency Response Institute
All rights reserved. No part of this publication may be reproduced, stored in a retrieval system, or transmitted in any form or by any means, electronic, mechanical, photocopying, recording, or otherwise, without the prior written permission of the publisher.

Outdoor Empire Publishing, Inc.
Printed in the United States of America
**Library of Congress Catalog #80-70906
ISBN 0-913724-24-6**

EMERGENCY RESPONSE INSTITUTE

9035 GOLDEN GIVEN RD., TACOMA, WASHINGTON

Phone (206) 531-3156

Dear Reader,

ERI is a progressive thinking organization composed of diversified emergency response management professionals, who share a common goal. That goal is to make the citizens of this nation more self-reliant and self-sufficient in their everyday lives, regardless of what emergencies or inconveniences they may face. We believe that to accomplish this goal it will be necessary to change habits, values and attitudes toward emergency preparedness: in essence, create an ethic. It is to this end that we are promoting emergency preparedness and response education for all factions of our society nationwide.

As the term Institute implies in our title, we are dedicated to advanced study, research and instruction in the specialized field of emergency preparedness and response. This publication was developed in consonance with many factions in the aviation community as an aid to the courses and workshops that ERI offers. You will find that even though it contains a wide diversity of subject areas from many sources, the material is oriented to general aviation. We believe that this contributes greatly to the book's uniqueness in the field of emergency preparedness and response.

Let us know if we may assist you in your efforts to promote general aviation safety and preparedness.

Robert "Skip" Stoffel
Patrick "Rick" LaValla
Co-Directors

Specialists in Emergency Preparedness Education

Acknowledgments

To Gene Fear, the Father of Emergency Preparedness Education, for his philosophy, advise, encouragement, support and critique.

To Dr. Paul Green for his continuing instructional assistance in our aviation survival seminars and for allowing us to borrow and "hitchhike" on his innovative concepts and approaches to emergency preparedness education.

To Bill Hamilton and Mac McIver of the Washington State Division of Aeronautics for their leadership in establishing programs in aviation emergency preparedness and survival, whose assistance and encouragement helped to produce this publication.

The following charts and illustrations were reprinted with permission: pg. 35, illustration adapted from a drawing by Gene Fear in "Surviving the Unexpected Wilderness Emergency" © 1973; pg. 42, U.S. Department of Commerce; pg. 76, Jim Mitchell, President Northwest Mountain Guides, Inc.; pg. 84, based on an illustration in "Death From Cold" by Marlin B. Kreider, Appalachia, June 1960; pg. 114, adapted from illustration AFM 64-3, U.S. Air Force; pg. 141, © "Physiology of Man in the Desert," E.F. Adolph and Associates, New York-London, Interscience Publishers, 1947.

Objectives

A Pilot Reading This Book Should

- React in the initial response phase of any emergency with a positive mental attitude toward effectively managing the situation.
- Know the most important lifesaving principles in making a successful emergency landing.
- Know the relationship between mental attitude, comfort zone and potential ability to function in the emergency environment.
- Know the priorities and necessities of life and their overall impact on effectively managing a survival situation.
- Be familiar with the most common reasons why pilots create their own emergency problems.
- Be familiar with the most common reasons why ELTs malfunction and how these problems can be prevented.
- Have a basic understanding of the skills necessary to be successful throughout the life support phase of any survival experience.
- Be familiar with the essential principles of Air Search and Rescue that will give a person the ability to effectively aid in their own rescue.

Table of Contents

Chapter I	**Yes, It Can Happen to You**......................9	

Chapter I   **Yes, It Can Happen to You**..................9
Place Yourself in this Situation • Another Example in a Different Environment • Common Sense Recommendations for Every Pilot

Chapter II   **Emergency Landing Techniques in Small Fixed-Wing Aircraft**..................18
Psychological Hazards • Techniques • Downed Aircraft Emergency Procedures Once on the Ground • Conclusion

Chapter III   **Establishing Life Priorities**..................29
What is Survival? • The Necessities of Life • The Whole Person Concept—The Necessities of Life in Concert with Mental Attitude

Chapter IV   **Physiological Considerations in the Emergency Environment**..................36
General Indicators of Major Body Upsets • Body Management Upsets—Cold, Heat, Altitude and Flying

Chapter V   **The Mental Aspects of Emergency Response and Survival**..................51
The Mechanics of Mental and Physical Response to Life Threatening Situations • Each Person Has an Individualized Comfort Zone • Controlling Fear • Each Person Acts in Accordance with Individual Beliefs

Chapter VI   **Emergency Preparedness Skills**..............60
Survival First Aid • Shelter • Checklists for Each Phase of the Emergency Environment • Clothing • Firecraft • Signalling • Improvising from the Aircraft • Cold Water Immersion Techniques • Determining Time and Direction • Water Procurement and Treatment • Food Procurement and Preparation

Chapter VII   **Some Considerations for Emergency Environments**..................129
Desert Environments • Winter and Arctic Type Environments • Salt and Fresh Water Environments • Sea Coastal Environments • Extreme Northern Latitude Environments • Important Survival Factors

Chapter VIII   **Emergency Preparedness Kits**..............133
Aircraft Emergency Kit • Supply Sources for Emergency Equipment

Chapter IX	**Search and Rescue**................................137	
	Search Verses Rescue • The National Search and Rescue Plan • The Search Effort • ELTs—Some Good Points and Some Bad • Flight into Foreign Countries • Conclusion	
Chapter X	**Managing Your Emergency in Hot or Cold Environments**..................155	
	Considerations for Managing an Emergency in Hot Environments • Considerations for Managing an Emergency in Cold Environments	
	Bibliography 160	

CHAPTER I
Yes, It Can Happen to You!

Post crash survival, in terms of coping with the environment and immediate physical needs, is usually thought of as a problem encountered only by pilots who fly long distances over unsettled, unfamiliar areas: deserts, open water, mountains, etc. This is not always the case. Numerous accidents have occurred within sight and sound of civilization. Many survivors who were unable to make their distress location known to others, suffered severely or died before rescuers could find them. An emergency preparedness (survival) kit on board the aircraft plus survival knowledge will often make the difference between life and death.

Place Yourself in This Situation

A startling example of what can happen to the pilot who takes for granted a flight over hot, barren country occurred in the semi-desert area of northeastern California. A 50-year-old Mooney aircraft owner, with mining interests in that region, had flown to a small airport near Hayden Hill, about 30 miles south of Alturas, early in the week. Having completed his business by Sunday afternoon, he took off and headed west toward his home in Yreka City. The fine September day offered perfect flying weather. His home field was less than an hour away, over high desert country that he had flown over so often that every butte and gulley seemed familiar. He did not file a flight plan, and he carried no survival equipment on board.

Would you file a flight plan?
☐ *Yes* ☐ *No*

Do you carry emergency equipment?
☐ *Yes* ☐ *No*

The pilot made a routine departure at 5:00 p.m. and climbed to his planned cruising altitude of 10,500 feet. As he began to lean out the mixture, the engine roughened. When he enriched the mixture, the engine stopped altogether. He had taken off with the fuel selector directed to the left main tank, which indicated full. Now he switched to the right tank, but was unable to accomplish a restart. Unsuccessfully, he tried various throttle and mixture settings. (Official findings by the NTSB would later point to a misaligned fuel selector indicator plate as the probable cause of the accident. While the fuel selector was pointing to a full tank it was actually in the "off" position or on the empty aux tank.)

Have you recently checked your aircraft maintenance?
☐ *Yes* ☐ *No*

The ground was now coming up fast. He radioed a quick "Mayday" and set up the airplane for an emergency landing. The flight path was following a county road through a dried-out river valley over rugged terrain strewn with boulders and gulleys. His first choice for a landing site was a straight stretch of the road. But he could see enough Sunday automobile drivers to increase the possibility that his landing would cause

Would your emergency thought processes be similar?
☐ *Yes* ☐ *No*

9

a collision. So he abandoned that thought. Next, he spotted a fairly level drywash, within gliding range, and headed for it. The cleared area was rimmed with several tall, gaunt pine trees, and strain as he might, the pilot was not quite able to clear them. His left wing struck two trees seven feet from the crown and shattered. The aircraft caromed off into another tree before crashing to the ground with a section of the tree imbedded in the engine and cabin. The pilot was rendered unconscious by the first impact.

When the pilot regained his senses, it was already growing dark. Evidently, he had been unconscious for an hour or more. He found that he had a badly broken leg and a cut on his head that had been bleeding over his face. He was unable to open one eye, and he was in quite a bit of pain. When he tried to move, he discovered he was pinned in his seat. The crash impact had jammed part of the panel and the right control yoke across the right seat, between the injured pilot and the only door. It was a discouraging situation—one that he did not feel up to facing that night, considering his injuries and throbbing head. He decided he could do nothing until morning except call periodically for help over the radio, although he was not sure whether it was working or not (Note: The radio, in fact, was **not** working.) He did not have an emergency locator transmitter (ELT) in the aircraft.

As night fell, the temperature on the desert plateau dropped down toward the freezing point. The injured pilot had only his lightweight flying jacket to huddle in. All night he remained in his cramped seat, cold and in pain, alternately drifting between semi-consciousness and wakeful suffering, waiting for the light and warmth of the next day.

When morning finally came he surveyed his situation. He could not move his right leg. It felt numb, alarming him considerably. Obviously he needed medical attention as soon as possible. Though he was not certain of his exact location, he had the impression that a small town was close by. Apparently no one had seen him go down. His only hope of attracting attention was to get out of the airplane, and that appeared to be no simple task. Since he could not move the obstructions between himself and the door he would have to dismantle some of them, at least the righthand control yoke, which was impossible without tools.

As he thought about his predicament he heard the sounds of cars passing along a road. He shouted and beat on the side of the airplane as loudly as he could in his weakened condition. There was no response, and he had hardly expected any. He also began to hear sounds of human activity as though some town nearby was stirring into life. The thought of being within earshot of help and unable to move was infuriating.

Reflecting on his ironic situation he suddenly remembered that he did have a small tool kit in the pocket behind the front seat. With considerable effort, he managed to get his hands on it, and proceeded to remove the control yoke wheel. It was slow work. He tired quickly and rested often. After two hours only one nut remained. It, however, could not be budged and was "frozen" in place with rust. He repeatedly strained at it

---

Is your aircraft equipped with a working ELT?
☐ Yes ☐ No

Do you carry emergency gear, including clothing appropriate for the area you are flying over?
☐ Yes ☐ No

Do you carry a first aid kit?
☐ Yes ☐ No

Is your aircraft equipped with signalling devices?
☐ Yes ☐ No

until he was exhausted and bloody-fingered. In between efforts to free the nut, he beat on the side of the airplane with his tools, shouted, and called on the radio. Still no response. He was getting hoarse and thirsty. He recalled that there was a can of soda pop under one of the seats. With much effort he twisted his body. His fingers finally touched a can. At first no matter how hard he tried, he couldn't get hold of it. Every move was a painful effort, but he kept trying. Finally he maneuvered the can into a position where he could pick it up, only to learn that it was not soda at all but a can of silicone spray lubricant.

Do you carry a small tool kit?
☐ Yes ☐ No

After his first flash of disappointment, he realized the value of his find and quickly sprayed lubricant on the frozen nut. After a short wait, he set the wrench on the nut again and felt it loosen. He soon had the wheel off, which enabled him to bend the seat forward far enough to allow him to crawl behind it, squeeze out the door, and move onto the wing. He was concerned about further injuring his broken leg and tried to support it with his belt as he moved.

It was now late Monday afternoon. It had taken him nearly a whole day just to free himself from the plane. He was certain that he had been missed by this time and that a search had been launched. He had heard, with great excitement, several planes fly over during the day, but the engine sounds remained faint and soon died away. Perhaps they had failed to spot him. He saw that he was in a partially wooded area, with no sign of life, or human habitation around. He was certain that he was not far from a road or town, but which direction? The weary pilot was determined to do something to help himself be found.

Would signalling devices have helped?
☐ Yes ☐ No

Negotiating the two foot drop from the wing to the ground was posing a serious problem for the injured man. When he found that it was impossible to slide off slowly, he simply let himself fall off. The impact was so painful that he passed out again.

When he regained consciousness, another night was setting in, and he realized that his chances of being found in the dark were almost nil. Unable to rise to his feet, he crawled along the ground, propelling himself with his forearms slowly and painfully. His goal was to crawl far enough away from the airplane to safely start a fire. His pockets yielded a total of two matches, and he gathered dry twigs into a pile.

Is there firestarting equipment on your aircraft?
☐ Yes ☐ No

Using punk from a rotted log, and with great care, he succeeded with the first match. The fire flared up rapidly, and moments later it was threatening to spread to the airplane. He had to beat the flames down with his jacket. This meant abandoning hope of attracting rescue that night, but he was too weary to mind the fire and too worried about the plane blowing up to leave a fire unattended. He then decided to sleep. The next morning he would crawl toward what appeared to be a peaked roof of a cabin, and he had hopes of finding help there. He fell asleep under the trees in a skimpy shelter improvised from leaves, tree limbs, and grass.

Would you prepare a fire circle?
☐ Yes ☐ No

Would you be able to utilize the parts of the aircraft for shelter?
☐ Yes ☐ No

The next morning he began his slow tedious task. The hilly terrain prevented his moving in a straight line, and a new pro-

Do you know that eating, with no water, increases dehydration?
☐ Yes ☐ No

Would you inventory equipment on the first day?
☐ Yes ☐ No

Is it a good idea to let others know your trip plan?
☐ Yes ☐ No

Are you familiar with the Air Search and Rescue System?
☐ Yes ☐ No

Do you carry provisions?
☐ Yes ☐ No

Would you do this on the first day?
☐ Yes ☐ No

blem was the effect of the sun. In his dehydrated condition he found that even a few minutes' exposure to full sunlight literally knocked him out. The only solution was to travel around any clearing and stick to the more difficult, wooded areas.

The top of the hill, where he had hoped to reach the end of his ordeal, offered a crushing disillusionment. The "cabin" he had been seeking turned out to be two tall, oblique, dead trees, leaning against each other in cathedral-like silhouette. Nowhere was there a sign of human life.

He realized that he could scarcely hope to reach the town wherever it might be, and that his best chance for survival, admittedly poor, was with the airplane. So he started back. Crawling back he found a few prickly pears, which he peeled and chewed. They seemed to help a little. By nightfall, he was back at the airplane and managed, finally, to retrieve his overnight case from the luggage area. Inside, there was aspirin to ease his pain and toothpaste which he used to moisten his lips. The only liquid he found in the airplane was wax, and he tried a swallow, experimentally, but had to spit it out. Rescue, he concluded, was out of his hands. He set up a shelter under the airplane wing and bedded down there, feeling increasingly weak and dehydrated.

Search and rescue efforts had been delayed nearly 36 hours, owing to uncertainty as to the missing pilot's intentions. When he failed to show up for supper Sunday night, his wife telephoned his business associates in Hayden Hill. They speculated that her husband had gone to Alturas, where he had mentioned having some matters to settle, and would probably spend the night there. It was only after the second night had passed without word from him that the FAA's Air Traffic Service was alerted and a search was officially initiated.

Over the next three days dozens of search flights were conducted by volunteer pilots and sheriff's helicopters. But since the pilot had reportedly flown towards Alturas the search was largely confined to the area northeast of Hayden Hill, which includes vast wilderness areas covering thousands of acres of extremely rough country. If a pilot actually did go down in this area and had no means of signalling rescuers or indicating the approximate crash site, it could be weeks, or months, or even years before he would be found.

But, of course, the pilot was not there at all. He had gone down some 30 miles **northwest** of Hayden Hill. During these three days he huddled in the shadow of his wrecked plane, scarcely moving, conserving his strength and chewing on whatever prickly pears he could find close at hand, his eyes and ears scanning for the rescue from the sky which was never to come.

By Thursday night he was so weak and dehydrated that he was convinced that he would not live through another day. He decided to make a last ditch effort to attract help. In the morning he would gather up whatever he could find in the way of clothes and rags, fasten them to his belt, and drag them into a clearing. There, he would lay out a big "X" and hope that it would be seen from the air. He realized that exposing himself to that much sunlight might well bring on a final state of unconsciousness (he was having intermittent blackouts most of the time), but it seemed to be his only chance. He hoped he could manage to struggle as far as the clearing. In his weakened condition, a hundred yards seemed like a hundred miles.

Soon after daybreak on Friday morning he began, slowly and painfully, putting together whatever suitable material for the signal that he could find. He was only barely conscious, and when he heard an engine and caught sight of a pickup truck glinting through the trees he was not sure whether it was real or an illusion. He banged on the airplane wing with all of his remaining strength. The appearance of two high school boys approaching with startled faces was the most welcome sight of the pilot's life. The youths, who had played hooky from school that morning to get in a little pre-season rifle target practice, drove back to town (less than a mile away) to notify the sheriff, who quickly arrived on the scene with an ambulance.

A month later the pilot was discharged from the hospital, but his ordeal was not quite over. He has had three operations on his leg to date, and he will need several more, to repair the shattered knee, before he can regain use of the limb. He knows that he was fortunate to have survived at all and that he contributed to his near demise by failing to file a flight plan or to notify anyone of his intended flight. But, he also contributed to his own survival by the courage and stamina he displayed in those five days of pain and despair. He does not know if he will ever be able to fly again, but if he does you can bet that he will have water stored on board the airplane, and a survival kit on hand. He knows, the hard way, that no pilot is immune to misfortune, and in the wilderness only those who are prepared are likely to survive.

Is a positive mental attitude necessary for survival?
☐ Yes ☐ No

The irony of this accident is that the pilot was never more than a mile from a town and a road, but was just as helpless as a man lost in a vast desert. This illustrates the unpredictable nature of aircraft accidents. Conditions are never exactly the same, but a basic knowledge of self-help procedures will increase your chances for survival in any environment.

Can you accept the fact that this could happen to you?
☐ Yes ☐ No

These are good reasons for staying close to the airplane if you go down on inhospitable terrain. Battered as it may be, the plane may afford the only shelter from the sun or other elements for miles around. And unless you are certain about your location you could exhaust and dehydrate yourself in a few hours by wandering around in the open. A sad case in point involved a 20-year-old student pilot who took off one day in May for his first cross country solo flight from San Diego to El Centro, California. He got off course, missed his destination and continued to fly until he ran out of fuel. He landed the aircraft on a deserted mesa near San Rosario, Mexico. Mistakenly believing that he was near the Salton Sea in southeastern California, he left a note stating that he was going to "walk to the shore ten miles away and turn north." When he found only barren cliffs before him, he struggled back to the airplane and left a second note declaring his intention to walk due north. He had no water. The pilot was found ten days later by a rescue team, dead from exposure.

Would you stay with the aircraft?
☐ Yes ☐ No

Would you conserve body water and energy by waiting for rescuers?
☐ Yes ☐ No

Another Example in a Different Environment

The capacity of enduring hours or even days in a snowbound environment is something winter pilots—especially those bound for ski resorts in the mountains—must consider before setting out. The following is an account of a pilot who, except for a combination of exceptionally fortunate circumstances, would not have lived to tell his tale. This particular incident occurred before ELT's were required.

The 26-year-old pilot was flying solo in a Mooney Mark 21 on a business trip from Salt Lake City to Portland, Oregon, some 640 miles. On that chilly first day of March he departed Salt Lake City at 1:45 p.m. The weather was VFR on departure, and he activated a VFR flight plan to the Oregon International Airport at Portland.

Would you file a flight plan?
☐ Yes ☐ No

An AIRMET (special weather report for light aircraft) issued that morning had warned that the mountains in Washington and Oregon would be mostly obscured, with occasional icing above 3,000 feet and occasional moderate turbulence. As a precaution, the Mooney pilot asked Salt Lake Departure Control for Radar Following Service to Oregon, but the Salt Lake tower was obliged to terminate the service at only 40 miles from the airport.

Do you know what kinds of winter gear should be carried?
☐ Yes ☐ No

The pilot had no difficulty maintaining VFR as he crossed from northern Utah and Nevada into Oregon. But now between him and Portland lay the Oregon Cascade Mountains, with peaks in the 10,000 to 12,000 foot range. He had to climb continuously to stay clear of clouds, and ultimately reached 14,500 feet, where he was experiencing strong headwinds. By this time he was significantly behind schedule and beginning to run low on fuel. One tank had already run dry, leaving him an estimated reserve of 25 minutes after reaching the Portland International Airport. To extend his reserve, at a point which he calculated to be 90 miles east of Portland, he reduced power and set up a descent rate of 300 feet per minute.

Would you consider an alternate airport?
☐ Yes ☐ No

Ninety miles east of Portland would have placed the aircraft somewhere east of the Deschutes River, which meant that Mount Hood (11,245 feet) still had to be flown over or circumnavigated. It was now after 6:00 p.m. and beginning to get quite dark, with ominous, towering cloud formations over the area.

At 7:10 p.m. Portland Approach Control was contacted by the pilot, who then gave his position as 45 DME miles east of the airport. Approach control was unable to make radar contact, presumably because of the close proximity of Mount Hood to the aircraft. The pilot indicated that he was on a direct course from the Kimberly VOR, which would bring him very close to the mountain. Portland Approach Control gave him the altimeter setting and suggested that he contact Seattle Center, whose radar might be able to pick him up.

The pilot called Seattle Center and gave his transponder code, his DME reading and the radial from Newburg VORTAC. He reported his altitude as 7,500 and descending.

"... Mooney two eight four X-ray, you should be close to Mt. Hood. Can you see the mountain?"

"Negative on the mountain. I've never flown this way before."

Because of the proximity of Mt. Hood which he could not see surrounded by clouds and darkness, the pilot decided to climb back out above the weather. He started a right climbing turn, found himself in heavy turbulence, and inadvertently penetrated a dense cloud formation. That was the last thing he remembered before he lost consciousness.

What had appeared to be a cloud was in fact the snow-covered side of Mt. Hood, which the aircraft struck at the 7,300 foot level under full power. The unexpected impact normally would have precluded survival. Winter conditions, at nightfall, would rule out any effective rescue effort until morning—weather permitting.

Luck played an important role. The aircraft was saved from immediate destruction by the heavy mantle of snow on the mountain, and by the fact that it happened to strike the slope at a shallow angle. It then had skipped along over a ridge of soft snow which slowed it down until finally flipping and burying itself in a deep pocket of fresh snow. It was the pilot's good fortune to have impacted above the tree line, at an angle which offered no solid resistance to the airplane.

As it was, the final impact did crush the forward section of the aircraft's cabin, but here again luck spared the pilot from serious injury. A short time before the crash he had released his seatbelt in order to reach back for a new chart in his flight bag. He had neglected to refasten the belt. Such an act can normally be expected to greatly increase the odds against crash survival. By a bizarre stroke of luck, he was thrown on impact into the rear of the cabin, which remained structurally intact. He was knocked out by a blow to his head, and remained unconscious for nearly 12 hours, with no protective clothing as the temperature fell below zero degrees.

Remarkably, at daylight the pilot was still alive. Although, the mass of snow around the wreckage had insulated him from the subfreezing weather, he was suffering from severe frostbite on his hands and feet.

The aircraft, based in southern California, carried no cold weather clothing or cold weather preparedness equipment. The pilot did have a suitcase with a change of clothing which he put on before climbing out of the wreckage.

The intense cold enveloped him immediately, and he sank to his waist in the snow. His head ached and he was confused about his actual situation.

Looking down toward the timberline, he spotted what appeared to be a hut, and began struggling through the deep snow in that direction. After a tiring descent of about a quarter of a mile, he found himself on a ledge that was too steep for him to climb down. He sat down on a stone outcropping. He could feel the cold numbness that was extending through his body. Death was not far away.

Once again luck was favorable. Although Mt. Hood had been hidden by clouds continuously the preceeding month, it

Is it a good idea to obtain complete terrain/environmental information before starting out?
☐ Yes ☐ No

Do you wear protective clothing when flying over cold environments?
☐Yes ☐No

Do you carry cold weather emergency preparedness equipment?
☐Yes ☐No

Do you know that sitting on cold surfaces increases body heat loss through conduction?
☐Yes ☐No

Do you know the principles of building snow shelters?
☐ Yes ☐ No

Do you know how to properly signal in a snow environment?
☐ Yes ☐ No

Would you have stayed with the aircraft?
☐ Yes ☐ No

was bright and clear the morning after the crash. This meant the search and rescue units, alerted during the night by the Seattle Center, were able to launch a search effort at first light. In less than an hour, a Civil Air Patrol airplane spotted the pilot's tracks on the snow which led them to the pilot, who was frantically waving his jacket.

Because it was impossible to safely land a helicopter on the ledge where the survivor stood, a pararescue team parachuted from an Air Force Reserve aircraft to administer first aid and to prepare him for hoisting on board a helicopter. An hour later he was in the emergency room of the Willamette Falls Hospital in Oregon City, and treated for severe frostbite, facial cuts, bruises, and a broken ankle. He had also lost a few teeth, but still managed a grin. He was lucky to be alive and knew it. Numerous winter crashes have occurred on Mt. Hood, and very few pilots or passengers have walked away from them.

Experienced pilots who fly in and around mountains know better than to rely upon luck. Winter and mountain flying can be safe provided that the pilot adheres to special flying practices appropriate for the terrain and weather.

In cold weather and in remote areas, the time it takes for search and rescue response can be critical for survivors. Survival time is increased by following proper flight procedures, knowledge, adequate preplanning and emergency equipment.

Edited from the FAA pamphlet **Survival In The Sun, Survival In The Snow.**

Common Sense Recommendations for Every Pilot

☐ File a flight plan of your intentions on every flight. It is the cheapest insurance policy you will ever be able to get.

☐ Insure the information on your flight plan is as complete and accurate as possible. Specify your route, alternate plans, alternative routes or stop-overs if weather is a problem. **The Search Coordinator will use this information to conduct any initial search effort.** The more information you give the better chance they have of finding you quickly.

☐ Close your flight plan within the time period required or notify the proper authorities to do so. You may become the object of a very expensive search. **It goes without saying that authorities will be upset at any pilot who does not close a flight plan and creates an unnecessary search.**

☐ Check the complete weather picture for your trip. Lack of good pre-flight weather briefings or complete disregard of these briefings is a predominant factor in many aircraft accidents.

- ☐ Carry recommended emergency and survival equipment during all flights. There is little point in being able to execute a beautiful forced landing and then dying of exposure or some other environmentally caused complication.
- ☐ Insure that your aircraft color scheme would assist in its visibility from the air. In too many cases the aircraft being sought is green and white and the area it is missing in is heavily timbered with evergreens and there is snow on the ground.
- ☐ If you become lost or disoriented in an area that is sparsely populated and are unable to communicate with any center, circle any residence or location of possible habitation at a safe low level until you are sure that someone has spotted you. If you press on and subsequently go down, you chances of speedy rescue should be enhanced by the observer.
- ☐ If flying VRF, do not make large deviations from your flight route to circumnavigate an unknown weather system. Do not attempt to fly through clouds unless you hold an instrument rating. Countless searches have been conducted for inexperienced pilots who have pressed on, become hopelessly lost and force landed or crashed in an area so remote from their intended route that hope for timely rescue has been all but impossible.
- ☐ Statistics prove that many downed pilots have simply not aided in their own rescue. Well placed, timely signals which contrast with the environment have been responsible for many successful rescues.
- ☐ Stay with your aircraft unless you are absolutely POSITIVE that help or civilization is close at hand, and you are capable of walking all the way out for help.
- ☐ Be positive in your approach to the situation. Even if you fail at something, you will have made some progress toward your ultimate goal, and that is to keep on living no matter what happens.

CHAPTER II

Emergency Landing Techniques in Small Fixed-Wing Aircraft

Any downed aircraft environment should be considered a hostile one. The passage of time does not work in your favor. Successful rescue in the shortest possible time is essential.

The factor of fundamental importance in rescue is assuring that the response system receives a distress alert in a timely fashion. Every pilot must have established emergency procedures. Several options open to the pilot at altitude quickly vanish in the down hill transition to the ground environment. Contact with flight centers or other aircraft, ELT activation etc., may not be possible on the ground.

Interest in seeking to learn "how to survive" is as commendable as thoroughness in checking the condition of the aircraft before taking off. Forethought and ability to make realistic assessment of dangers should be balanced with the knowledge and the skill to survive emergencies. The ability to survive requires good management practices. Allowing yourself to be placed in a survival situation—except under exceptional circumstances—is a reflection of poor management.

An aviation emergency presents a special survival situation which begins before a person reaches the ground. The successful transition from flight to the ground is the critical part of the emergency. National statistics indicate 25 percent of general aviation accidents involve emergency landings. Guidelines show almost any terrain can be considered suitable for a survival crash landing if the pilot uses the aircraft structure for protection. Much of the following has been condensed from the National Transportation Safety Board Report No. NTSD-AA-72-3, "Emergency Landing Techniques in Small Fixed-Wing Aircraft."

Psychological Hazards

Several factors may hinder a pilot's ability to act promptly and properly in an emergency.

Reluctance to Accept the Emergency Situation

A pilot who allows the mind to become paralyzed at the thought that the aircraft will soon be on the ground no matter what is done severely handicaps handling the emergency. Unconscious desire to delay the dreaded moment may cause such errors as failure to lower nose to maintain flying speed; failure to lower collective to maintain rotor rpm (in helicopters); delay in selecting the most suitable touchdown area within reach; and, general indecision. Desperate attempts to correct whatever went wrong, at the expense of aircraft control, fall into the same category.

Desire to Save the Aircraft

A pilot conditioned by training to expect a relatively safe landing area whenever the instructor closed the throttle for a simulated forced landing, may ignore basic rules of airmanship to avoid a touchdown in terrain where aircraft damage is unavoidable. Typical consequences: making a 180° turn back to the runway when altitude is insufficient; stretching glide without regard for minimum control speed in order to get into a better-looking field; accepting an approach and touchdown situation that leaves no margin for error. Trying to save the aircraft, regardless of risks involved, may be influenced by two other factors: pilot's financial stake in the aircraft; and, certainly that an undamaged aircraft implies no injuries. There are times when a pilot should be more interested in sacrificing the aircraft for safety of passengers.

Undue Concern About Getting Hurt

Fear is a vital part of our self-preservation mechanism. However, when fear leads to panic, it invites what we most want to avoid. A pilot who allows choices in selecting a touchdown point for a fully controlled crash has no reason to despair. Survival records favor those who maintain composure and know how to apply the general emergency concepts and techniques developed throughout the years.

In summarizing the role of psychological hazards it appears the success of an emergency landing under adverse conditions is as much a matter of mind as of skills.

Techniques

The "school solution" to a forced landing emergency requires the following actions:
- Maintain aircraft control (establish a glide at proper speed).
- Select a field and plan an approach.

These actions may be combined with attempts to correct the emergency, especially when the pilot knows what went wrong (carburetor heat, mixture, fuel selector, etc.). Trouble-shoot the cause of the emergency only on a time-available basis. It may be a full-time job just controlling the aircraft. Losing one engine of a light-twin during the critical takeoff phase may give the pilot only a split second to decide what is best: relying on performance charts, or obeying impulse to reduce power on the good engine to maintain control. Each pilot should determine minimum altitude at which an attempt would be made to turn back to the runway if the engine failed on takeoff. Experimenting at a stage altitude will let you know about how much altitude is lost in a decending, 180° turn at idle power. Adding a safety factor of about 25 percent gives a practical "decision height." The ability to make "180 degrees" does not necessarily mean you can get back to the runway in a power-off glide. This depends on wind, height reached, distance traveled during climb, and glide distance without power.

Terrain Selection

A pilot's choice of emergency landing sites is governed by the route selected during preflight planning, height above the ground when the emergency occurs, and airspeed (excess airspeed can be converted into distance and/or altitude).

The only time the pilot has a limited choice is during low-and-slow portion of take-off; under those conditions changing heading only a few degrees may insure a survivable crash.

When beyond gliding distance of a suitable open area, the pilot should judge terrain for energy-absorbing capability. If the emergency starts at considerable height above ground, select desired general area rather than specific spot. Terrain appearance from altitude can be misleading and considerable altitude lost before the best spot is pinpointed. As a rule, do not change your mind more than once; a well-executed crash landing in bad terrain can be less hazardous than an uncontrolled touchdown on an established field, although bad plans should be discarded for a better one.

Aircraft Configuration

Since flaps improve slow speed maneuverability and lower stalling speed, use them during final approach when time and circumstances permit. However, the increase in drag and decrease in gliding distance call for caution; premature use of flap and lost altitude may jeopardize an otherwise sound plan.

There is no hard-and-fast rule for the best position of a retractable landing gear at touchdown. In rugged terrain and trees, or during impacts at a high sink rate, an extended gear would definitely help protect the cockpit/cabin area. However, a collapsing gear has side effects such as a ruptured fuel tank. Manufacturer's instructions—if given—should be followed.

When a normal touchdown with ample stopping distance is assured, a gear-up landing on level but soft terrain, or across a plowed field, may cause less aircraft damage.

Shutting off the aircraft's electrical system before touchdown reduces the likelihood of a post-crash fire. However, the battery master switch should not be turned off while the pilot needs electrical power to operate vital systems (flaps, hydraulics, etc.). Positive aircraft control during the final part of the approach has priority over all other considerations, including aircraft configuration and cockpit checks. Use any power available from an engine running irregularly. However, to avoid unpleasant surprises, switch the engine and fuel off just before touchdown to insure control over the situation. A cooled-down engine reduces the fire hazard considerably.

Approach

When there is time to maneuver, approach planning should be governed by three factors: wind direction and velocity, dimensions and slope of the chosen field, and, obstacles in the final approach path.

These factors are seldom compatible. When compromises have to be made, aim for a wind/obstacle/terrain combination that permits a final approach with some margin for error in judgment or technique.

A pilot who overestimates gliding range may try to stretch the glide across trees, powerlines, etc. It is sometimes better to plan the approach over an unobstructed area, regardless of wind direction. Hitting obstacles at the end of a ground roll, or slide, is much less hazardous than running into something at flying speed before touchdown.

There are no specific rules for the pattern to be flown; there may not even be time to set up a pattern. It is important to get into position so that the selected spot can be reached using normal techniques such as playing the final turn (turning in early or late, depending on altitude), slipping, and moderate S-turns. If altitude must be lost over or near the chosen field, keep the field within gliding distance. Speed control is vital during all maneuvers.

Optical Illusions in the Rain

Very serious errors in vision can result during rainy weather flight conditions. The problem goes beyond poor visibility and introduces a refraction error that can be serious. Because of the eye response to bright upper part and dark lower part in the sight picture, and the reduced transparency of the windshield, a pilot may think the horizon is actually lower than it is. In addition, the ripples on a steeply slanted windshield cause objects to appear lower than they are. The error may be a result of either of these two reasons or a cumulative effect of both. The result may be an approximate 5 degree angle or about 200 feet in a half mile.

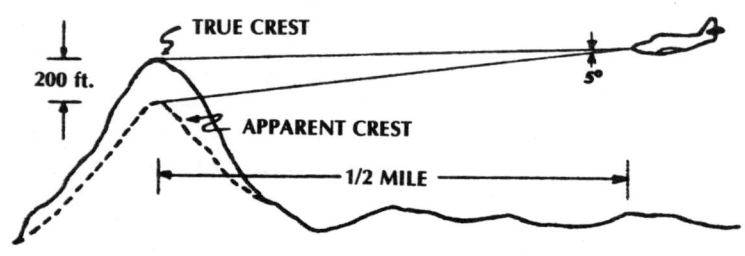

One of the hazards of flight in rain.

This problem becomes extremely critical in mountainous terrain and on final approach. Precautions should be taken to allow for this refraction error. Tests have shown that rain repellents on the windshield have a significant effect on visibility.

Crash Safety

A pilot faced with an emergency landing in terrain where extensive aircraft damage is inevitable should keep in mind that avoiding crash injuries is largely a matter of keeping vital structure (cockpit/cabin area) relatively intact. This is accomplished by using disposable structure (wings, landing gear, fuselage, bottoms, etc.) to absorb the violence of stopping the plane before it affects occupants, avoiding forceful bodily contact with interior structure.

There are many things that passengers can do to reduce injurious contact with the interior of the cockpit. Using small pillows, extra clothing and proper seat position can all decrease injuries. It may well be wise to unlock the door and wedge something in it to insure the ability to exit. Removal and securing of items above and behind the rear seats will eliminate flying projectiles. Common sense should be used as guidelines along with the realization that everyone in the aircraft has a responsibility to aid in survival activities.

Energy Absorption

The advantages of sacrificing dispensable structure are demonstrated daily on highways; a head-on car impact against a tree at 20 mph is less hazardous for a properly restrained driver than a similar impact against the driver's door. Accident experience indicates the amount of crushable structure between occupants and the part of the aircraft hitting the ground has a direct bearing on how much crash force occupants receive and, therefore, effects their survival.

Dispensable aircraft structure is not the only way to absorb energy in an emergency. Vegetation, trees, and manmade structures, may be used. Cultivated fields with dense crops such as mature corn and grain, are almost as good as bringing an aircraft to a stop with repairable damage as an emergency arresting device on a runway. Brush and small trees provide considerable cushioning and braking without destroying the aircraft. When crash landing into natural and manmade obstacles stronger than the dispensable aircraft structure, the pilot has to plan the touchdown so that only nonessential structure is "used up" to slow the plane down.

Occupant Restraint

The second requirement—avoiding forcible contact with interior structure—is a matter of wearing seatbelt and shoulder harness. If the occupant is not strapped in, that person will not benefit from keeping the cabin relatively intact, but will bounce around inside, a so-called second collision. Wearing only a seatbelt permits the body to jacknife and hit objects in the cabin. This has caused fatalities in otherwise survivable aircraft accidents. Since so few light aircraft are equipped with shoulder harnesses, the pilot should try to avoid nose-first impact against solid obstacles. Pilots should insist on the routine use of shoulder harnesses if the airplane is equipped with them.

Simulated forced landings occasionally become real at high sink rate when the engine fails to respond as anticipated. Automatically raising the nose when the throttle is advanced for a go-around, without waiting for engine acceleration, can lead to destructive sink rates. The pilot should maintain proper approach speed and attitude until the engine responds; this also applies to go-arounds from balked landings.

Touchdown

The importance of controlling the aircraft's attitude and sink rate at touchdown has already been explained. Since an emergency landing on suitable terrain resembles a normal landing, only the more unusual situations will be discussed.

Confined Areas

When the intended touchdown point is reached and the remaining open space is very limited, it may be better to force the aircraft down on the ground than to delay touchdown until it stalls (settles). An aircraft decelerates faster on the ground than while airborne. It may be desirable to ground-loop or retract the landing gear in certain conditions. A river or creek can be inviting in otherwise rugged terrain, if the water or creekbed level can be reached without snagging the wings. The same concept applies to road landings with one additional caution; manmade obstacles on either side of a road may not be visible until the final approach. Road traffic has priority. When planning the approach across a road, remember that most highways, and even rural dirt roads, are paralleled by power or telephone lines, and keep a sharp lookout for supporting structures or poles.

Trees

Although a tree landing is not attractive, the following general guidelines will help to make it survivable.

- Use the normal landing configuration (full flaps, gear down).
- Keep the groundspeed low by heading into the wind.
- Make contact at minimum indicated airspeed, but not below stall speed and "hang" the aircraft in the tree branches in a nose-high landing attitude. The underside of the fuselage and both wings provide more even and positive cushioning while preventing penetration of the windshield.
- Avoid direct fuselage contact with heavy tree trunks.
- Low, closely-spaced trees with wide, dense crowns (branches) close to the ground are much better than tall trees with thin tops which allow too much free-fall height. (A free-fall from 75 feet results in an impact speed of about 40 knots, or 4,000 feet per minute.)
- Ideally, both wings should meet equal resistance in the tree branches to maintain proper aircraft attitude; it may also preclude losing one wing, which leads to a more rapid and less predictable descent.
- Always aim for the softest and, when possible, the lowest part of a tree or tree line. Judge trees by their ability to slow the aircraft's forward speed the same way a firefighter's safety net catches people.
- If heavy tree trunk contact is unavoidable once the aircraft is on the ground, try to direct the aircraft between two properly spaced trees to involve both wings at the same time. Don't attempt this "maneuver" while still airborne, as recommended in some textbooks.

Mountainous Terrain

The variety and irregularity of mountainous terrain makes it impossible to list general rules. Learn to instinctively avoid situations where an emergency leaves no choice; for instance, flying needlessly low and slow over cragged terrain.

In mountainous terrain, a short glide may bring the aircraft over lower lying terrain, increasing effective altitude and terrain choice; maintaining a comfortable cruise speed assures this advantage.

Make landings upslope whenever possible, depending on the terrain at the end of the slope and avoiding a situation where an excessive roll or slide would bring the aircraft to a sharp dropoff. When landing on a pronounced upslope, maintain enough speed to change the aircraft's descending flight path, just before touchdown, into a climbing one that approximately parallels the slope.

Water (Ditching)

A well-executed water landing probably involves less deceleration violence than a poor tree landing or a touchdown on extremely rough terrain. The reason for the apparent reluctance of some pilots "to take to the water" when there are no suitable alternatives may be the certainty of losing the aircraft or the fear of getting trapped. Actually, a fixed-wing aircraft ditched at minimum speed and in a normal landing attitude will not sink like a rock upon touchdown. Intact wings and fuel tanks (especially when empty) provide flotation for several minutes even if the cockpit may be just below the waterline in a high-wing aircraft.

When considering the feasibility of ditching, take the following factors into account:

- The water temperature and the estimated time to be spent in the water. (The survival time in water with a temperature of 33° Fahrenheit is less than one hour for the average person.)
- The proximity to land.
- The physical conditon of the occupants and their ability to swim.
- The availability of lifevests and other water survival equipment.
- The number of occupants and the number of usable exits.

Loss of depth perception may occur when landing on a wide expanse of smooth water, with the risk of flying into the water if stalling-in from excessive altitude. To avoid this hazard, the aircraft should be "dragged in" when possible. Use no more than intermediate flaps on low-wing aircraft; the water resistance may break off one flap and slow the aircraft. Keep retractable gear up. Insist that all occupants keep their belts fastened until the aircraft has come to a complete stop; this insures impact protection and prevents confusion about exit locations regardless of aircraft attitude and light conditions. Ditching downstream in a swift-running river reduces the relative groundspeed.

Snow

Executing a landing in snow is like ditching, in the same configuration and with the same regard for loss of depth perception (white-out) in reduced visibility and on wide open terrain. An even snow layer, several feet thick, may blanket smaller obstructions and make otherwise rough terrain more suitable. Avoid pronounced "humps" that may hide larger obstructions.

Downed Aircraft Emergency Procedures Once on the Ground

Survival following an airplane crash is unique. The crash is a traumatic experience and a good percentage of victims are injured on impact. Injury affects all aspects of survival and should be considered heavily in preplanning. In air crashes, energy levels are usually high and available for a longer period.

The immediate priority after a crash is to get clear of the aircraft as quickly as possible. This provides needed time for gasoline to evaporate, engines to cool, and possible electrical shorts to dissipate. During this interim, check for injuries, give first aid where necessary, and begin to formulate a plan for temporary shelter. There may be inclement weather to cope with as this is a leading cause of small aircraft accidents. (Note of caution: be careful in removing any casualties from the aircraft, particularly those with back or spinal injuries.)

Is the aircraft ELT working? Crash impact forces place a tremendous amount of G-forces on the transmitter (as much as 44,000 foot pounds). Usually inadequate mounting and improper positioning results in severed antennas, or complete expulsion from the aircraft. This device is one of the primary means of signalling distress and should be located immediately and checked for correct operation. Be certain the triggering mechanism has fired and antenna is properly attached and in place. Emergency antennas can be improvised from pencils, pens, wire, or other metal placed into contact with the receptacle. Victims must not overlook the possibility of using the aircraft radio at intervals by broadcasting on local frequency every 15 minutes during good weather. Morse Code SOS can be used if batteries are weak.

Remain with the aircraft if possible. It contains basic ingredients for warmth, shelter, and signalling materials. Keep it visible from the air. Clear brush, trees, snow, and other debris from surfaces to enhance contrast with environment and insure visibility. If possible, keep a fire burning and maintain hand signals within easy reach. (Flares, penguns, vari-pistols.)

Conclusion

A pilot who knows his aircraft and understands the what and why of techniques to insure a survivable emergency landing under adverse conditions has no reason for morbid fear of the possibility of being forced down. The peace of mind associated with this knowledge should improve the pilot's overall performance which, in turn, may prevent an emergency or substantially improve its outcome.

The Six Common Sense C's of an Emergency Situation

- **Confess:** Confess to yourself that you are lost or are experiencing some kind of difficulty.

- **Climb:** Gain as much altitude as the weather will permit to allow greater range on your radios.

- **Communicate:** Let others know of your distress. Broadcast on 121.5 or 243 Mhz.

- **Conserve:** Conserve by reducing the throttle setting to endurance optimum configuration.

- **Comply:** Follow explicit instructions from controlling agency.

- **Consult:** Consult **Survival Sense for Pilots** for additional options and instructions.

Message Format for Emergency and Distress Calls

- **MAYDAY, MAYDAY, MAYDAY. This is aircraft Cessna, Mooney, etc., three one two six romeo.**

- **Position or estimated position** (state which of the two). Try and be as precise as possible. Also state time.

- **Heading** (True or Magnetic).

- **True or estimated airspeed** (state which of the two).

- **Altitude**

- **How much estimated fuel on board or flying endurance** (in hours and minutes).

- **The nature of the emergency** (loss of oil pressure, extreme icing, etc.)

- **What are your intentions:** Ditching, emergency landing, continue on heading as long as possible.

- **Assistance desired:** Fix, bearing, position on radar, heading of nearest clear weather.

Repeat and Continue to Transmit if No Answer

CHAPTER III
Establishing Life Priorities

What is Survival?

Traditional definitions of survival imply primitive conditions, living off the land, or battling the elements to live through some precarious predicament in the wilderness. Depending on the perspective that individuals have, or the life activities sphere in which they operate, survival means different things to different people.

A survival situation is one in which an individual's very existence is threatened. Some kind of action is necessary to alleviate a threat to life. To survive is to continue existing, to live in spite of some event or adverse set of circumstances. In brief, survival is simply living a minute longer by any improvised means possible. The key term here is improvise. Seldom will an individual enter into a life threatening experience with no material resources.

The ability to survive is infinitely more complex than merely being able to "live off the land". Statistics prove that nearly all survival situations today are short term (72 hours or less). The much publicized long term survival experiences make up fewer than five percent of all incidents. In establishing an effective, rational response to any survival situation, it is imperative to realize that actions and needs must be prioritized. Preconceived ideas about survival can precipitate incorrect priorities and actions in the initial moments of stress. These false ideas may lead inexorably to a fatal end. First hand accounts and interviews substantiate the fact that the decisions made in the first hours or minutes of an emergency have a great bearing on the ultimate outcome.

It is essential to understand the human body; not entirely as a physician understands it, but how it reacts to or is affected by environmental stress. In short, a working knowledge of how to properly manage your body in an emergency environment is the key to survival. Proper body management is a skill in maintenance and control of your own essential (limited) body resources and problem solving capabilities. If mismanaged, body upsets will directly affect a person's ability to exist in any environment for any period of time.

Traditional survival skills have a definite place, but they must be tempered with an understanding of how humans function under mental, as well as environmental stress. An understanding of why the necessities of life must be prioritized can be gained by polling the multitudes of "survival experts" around the country. Most of these authorities express conflicting views as to what the priorities of life should be in a genuine survival emergency. Who is to say which authority is right? Every incident is different just as every individual who might be involved is different. However, the necessities of life do not change, and an analysis and understanding of their priority should always outline a general course of action.

Probably the greatest obstacle an individual will have to overcome is mental attitude. Since survival is a nearly 100 percent psychological challenge, the brain becomes the individual's most important tool. Recognizing the body's needs and acquiring those needs will always require an initial thought process based upon previous experience and education.

People live, work and play in a comfort zone. Just as people differ, so do comfort zones and their limits. All people are unique in their abilities. Since people are creatures of habit, and the technology they have created takes much of the unexpected out of their lives, most rarely go outside of their comfort zones. Experimental programs in high risk and challenge activities have proved conclusively that people who operate on the outskirts of their comfort zones and continually push on those limits, react better in the emergency environment. Under repeated stress situations, people learn to make more and more effective adjustments. By contrast, the typical emergency situation results in disorganized and frequently shock-like behavior. Unknown factors during any emergency will be numerous and variable, factors that fall outside of the comfort zone. Some compound a situation. At the same time, others are counteracting it. Rational, coherent thinking which recalls similar situations or circumstances from past education or experience will, in most cases, be the most effective method of gaining control of the situation.

Modern technology has accustomed people to light switch conveniences that eliminate thirst, hunger, cold, heat and, in most cases, fear. In an unexpected survival situation, confidence and self-reliance will play a key role in determining the ultimate outcome of that experience. Unfortunately, people are not only dependent upon their technology, but are products of a culture which has advanced beyond the necessity of being self-reliant on a daily basis. In short, people are dependent on their technology. Self-reliance is developed through actual experience and dependence on one's self for every day needs. There is no substitute for actually having performed a task which is designed to save lives.

The Necessities of Life

What does it actually take for a human being to stay alive for an indefinite period of time? Typical answers might include clothes, a house, a car and three meals a day. However, it is obvious that several thousand years ago humans did not have these conveniences, and still survived. By listing the necessities of life and asking the questions, how long can a human live without each item, it is then possible to prioritize these necessities in order of importance. The time factors involved with this type of analysis should be in terms of minutes, hours, days and weeks. Although this approach may seem rather elementary, it will provide a foundation for good judgement when dealing with all life threatening emergencies.

Air

In considering the basic body functions, breathing is essential to maintain life on a minute-by-minute basis. Therefore, air (specifically its life sustaining component, oxygen) is the first priority on the list. The average survival time without air ranges from three to five minutes.

Shelter

Inherent with a major portion of survival situations is having to cope with mother nature and inclement weather. Since the human being is designed to live naked only in areas where temperatures are very close to that of the body, environments outside this realm post many and varied body protection problems. Anything that protects the body can be called shelter. Clothing is shelter in close proximity to the body. The problem is to provide adequate shelter for a specific environmental situation. There are temperature extremes in the United States where inadequate shelter could cut survival time to a few short hours. In some cases, it could even be less than one hour. With this in mind, shelter is the second priority on the list. Fire and warmth can also be considered under shelter because, by definition, they protect the body in a cold environment.

Rest

As already mentioned, energy levels within the body are vitally important when looking at the ability to cope with any given situation. There is a good possibility that it may not be possible to renew or add to energy levels during an emergency. Therefore, it is imperative to conserve what is already present in the body. One of the most efficient ways of doing this is through adequate and prudent rest. The human body is a wonderful machine which is capable of doing many things if provided with proper care and maintenance. Proper and timely periods of rest in stress situations conserves energy for future use, and rids the body's tissues of CO_2, lactic acid and other body wastes. Taking a short time for rest will also provide time for reflection and analysis of the situation. Mental rest is just as important as physical rest, if not more so. Extreme stress periods (both physical and mental) with no rest can cut survival time to as little as 30 hours. With this in mind, rest is the third priority on the list.

Water

It is generally accepted that water and its associated problems, hypohydration and potable condition, are among the most pressing in any survival situation. The body is comprised of approximately 2/3 water which is stored within the tissues, circulatory system and internal organs. Water is essential for body temperature regulation, waste elimination and digestion of food. While at rest with no activities, the body utilizes approximately two quarts of water a day merely to carry on normal body functions. Consumption of any amounts less than this will result in gradual hypohydration. Activity of any kind drastically increases the body requirements for water. Therefore, it is not the water that should be rationed, but the sweat and activity. A 2.5 percent water deficit can occur in a few hours. This can result in a 25 percent loss of efficiency. Extreme conditions without water could cut survival time to three days or less if prudent methods of water conservation are not used. Water, then, is number four on the list of priorities for maintaining human life.

Food

Contrary to popular belief, the human body does not require three meals a day to remain alive. Countless people across the world today live on far less. Many have only one meal a day and some not even that. Records and statistics from survival experiences show numerous accounts of 40 to 70 day periods with no caloric intake. This fact coupled with current records showing over 98 percent of the survival situations arising today last 72 hours or less, emphasizes the fact that procurement of survival food should be last on the list of priorities.

Energy levels within the body play an important role in warmth and performance. It is a common fact that high energy snacks provide additional calories for heat production as well as movement. Although food would play a minor role under most circumstances, cold environments require many calories to maintain normal body temperature. As was mentioned before, conservation of the energy levels already in the body is one of the most important factors to keep in mind.

The Whole Person Concept — The Necessities of Life in Concert with Mental Attitude

So far, air, shelter, rest, water and food have been presented in what seems to be a logical order of importance. However, an often overlooked but vital priority is a mental attitude which is positive in nature and proper in its perspective for a given situation. This is usually the result of a thorough background in the knowledge and skills necessary to preserve human life in a hostile environment.

An extremely important part of emergency response form an individual's standpoint is the "whole person concept." An understanding of this basic theorem enables a person to better put into perspective the priorities of any emergency environment. The "whole person concept" as it applies to response is a term which ties both mental and physical body processes together. Very simply put, what affects a person physically, also affects mentally, and whatever affects mentally will ultimately affect physically. Realization of this fact alone can have tremendous impact on any person's ability to cope with stressful environments.

If all the other priorities have been maintained or procured, but an individual lacks the will to live or the ability to cope with the situation mentally because of physical condition, then all is lost. People are at the mercy of the elements and whatever luck might be thrown their way. The power of the mind must never be underestimated.

It is worth a note of caution here to emphasize that although a Positive Mental Attitude (PMA) should be number one on the list of priorities, it would be possible to take negative action even though your attitude was positive. Mental attitude must not be misconstrued as a cure-all for all survival situations. Positive Mental Attitude does very little to analyze the body's enemies, recognize a threat to life or understand the counteractions necessary in an emergency situation. It must be coupled in the majority of cases with the proper education and background.

In order to further clarify a PMA, it is helpful to break the concept down into two areas. The first, already discussed very briefly, is the "will to live". Expressed simply, it might be described as an overwhelming urge to survive no matter what the odds or circumstances. It is debated extensively as to whether or not humans have a natural instinct for preservation. Numerous accounts during World War II and the Korean conflict point out the fact that death seems to be the easiest way out during periods of extreme physical and mental stress. People have the power to will themselves to death. By the same token, they have the power to will themselves to live. The main focus of concern at this point deals with the problem of instilling this "will to live" throughout society.

The second area of a PMA deals specifically with the "whole person concept" and a person's problem solving ability. It is possible to be thoroughly knowledgeable, experienced, have a PMA commensurate with the situation, and still not operate at 100 percent of our capability.

Consider this following example. The human body operates at a optimum temperature of 99° Fahrenheit. Regardless of the environmental temperatures, there is a relatively narrow range in which the body's core functions are at peak efficiency. This also holds true for the peripheral areas of the body. For instance, in a cold environment as the body temperature begins to drop, the blood supply to the extremities is automatically reduced in order to conserve heat in the central core where the vital organs are located. Unfortunately, due to the reduced blood flow and lowered cell activity, electrical impulses from the peripheral areas are also reduced and very soon the brain begins to lose touch with reality. Irrational behavior coupled with reduced coordination often will precipitate a fatal accident. In the same respect, heat stress, hypohydration, chemical imbalances and a host of other physiological problems can drastically lower the body's operating efficiency. Knowledge, experience, and mental attitude working at 100 percent will be of no value if the body is only operating at 60 percent efficiency. Safeguard the problem-solving ability by having a thorough understanding of indicators within the body.

What this all boils down to is this: a PMA which is proper in its perspective for a given situation ultimately assumes the highest of all priorities during any survival situation. This is a product of experience, education, and training. Shelter is the next most important concern in the majority of survival emergencies. The body must be protected at all costs through a combination of recognizing indicators and improvising needs. Prudent conservation of existing resources which include water, energy, and heat is also paramount to a successful survival experience.

Necessities of Life in Priority Order

Requirement	Survival Time if not Fulfilled
1. Positive mental attitude	Depends entirely on you
2. Oxygen in air	3 to 6 minutes
3. Body shelter from extreme temperatures	3 to 4 hours
4. Rest	30 hours in extremes (mental as well as physical)
5. Water	3 days in extremes
6. Food	3 weeks or more

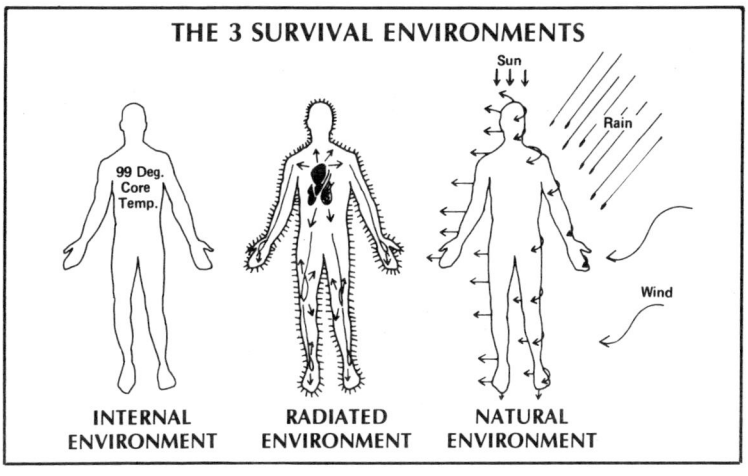

THE 3 SURVIVAL ENVIRONMENTS

INTERNAL ENVIRONMENT — RADIATED ENVIRONMENT — NATURAL ENVIRONMENT

Major Body Enemies in Order of Importance

The Enemy	How it Threatens
1. Your mind	Attitude, imagination, fear, panic
2. Temperature	Body core temperature must be maintained within a narrow 12-degree range for effective, rational behavior
3. Injury	May affect mobility, coordination, and ability for self-help
4. Disease or infection	Constant threat, normally held in check by the body's defense mechanisms

CHAPTER IV
Physiological Considerations in the Emergency Environment

It has been established that mental and physical functions of the body are virtually inseparable. Both can impact our ability to effectively deal with the emergency environment. Although injury will be the greatest and most immediate physical concern, environmental stresses, both at altitude and on the ground, can produce equally dangerous but more subtle, insidious results.

This chapter provides the pilot with only the essential factors relating to physiological complications. What are the potential problems? After establishing what can go wrong, it is necessary to understand what conditions are usually present which produce these problems. Although physical injury presents relatively standard first aid problems to solve, an isolated survival situation may dictate more definitive action. Environmental stresses which result in overt physical symptoms need to also be identified.

Signs and symptoms as well as treatment and ways to prevent these problems are also essential factors. The following set of matrix have been set up to eliminate nonessential material and present "need to know" information related to physiological and first aid problems. It is designed to be used as both a checklist in an emergency and a frequent review of first aid procedures.

General Indicators of Major Body Upsets

Cold Indicators

Self Analysis

Symptom	Significance
Mild shivering, mild skin pain.	Skin temperature lowering (body is losing heat faster than it is being produced).
Involuntary shivering, numbness, loss of dexterity, inability to move muscles.	Immediate action must be taken.

Observed in Others

Symptom	Significance
Mild shivering, hands in pockets, arms and legs held close together.	Body losing heat faster than it is producing it, and individual is unconsciously trying to conserve.
Clumsiness, general loss of coordination. Withdrawal from activities, poor articulation. Whitish or bluish color in ears, nose, hands, and feet.	Core temperature beginning to drop. Immediate action is necessary to rewarm individual.
Poor judgment, possible hallucinations.	Reaching critical temperature of body core.
White spots on skin.	Indication of frostbite.

Heat Indicators

Self Analysis

Symptom	Significance
Thirst, discomfort, easy fatigue.	Not acclimatized. Start of dehydration.
Extremely uncomfortable, sweating.	Body having trouble dissipating heat.
Muscle efficiency and capability slows down.	Body not coping with heat gain.
Excessive thirst.	Dehydration.
Painful voluntary muscle spasms.	Cramps due to chemical imbalance in blood and tissues.
Intense, frequent nausea.	Dehydration.

Observed in Others

Symptom	Significance
Pinkish to red skin.	Possible sunburn.
Sweat stains on clothing.	Losing too much water through perspiration.
Increased fatigue — frequent stops for rest.	Body not coping with heat.
Withdrawal from group, lethargic movements, possible hallucinations.	Dehydration. Body temperature rising above normal. Possibly close to critical temperatures.

Salt or Chemical Deficiency Indicators

Symptom	Significance
Cramps in major muscles, fatigue, intense weakness, headache, irritability.	Salt output has exceeded intake.

Dehydration Indicators

Symptom	Significance
Mouth	Dry, parched.
Headache	Usually an early sign of dehydration.
Skin	Slow to return to normal when pinched or pulled.
Urine	Dark, another early dehydration sign.
Thirst — may be intense.	Thirst alone is not a strong enough indicator to prevent dehydration. Could indicate other difficulties, but always accompanies dehydration.
Irritability, fatigue, nausea.	
Blurred eyesight, impaired judgment, loss of coordination, hallucinations.	Advanced stages of dehydration.

Effective Temperature at Which Common Cold Weather Injuries Occur (No Wind)

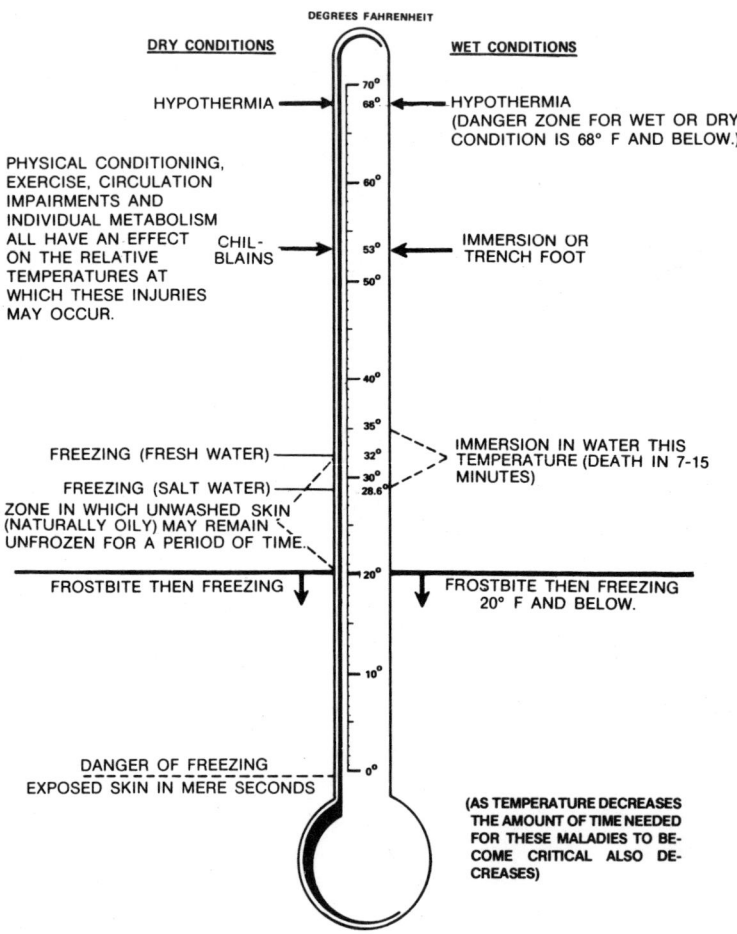

Body Upset	Predisposing Factors	Signs and Symptoms	Prevention	Treatment
Sunblindness (also snowblindness) See same topic in heat section for more details.	Bright sun reflecting on snow and ice. Occurs on overcast days (ultraviolet light).	See same topic in heat section.	Sun-goggles, sunglasses with side protection. Improvise glare shades.	See same topic in heat section.
Sunburn See same topic in heat section for more details.	Skin exposed to sun, and sun reflecting on snow and ice. Occurs on overcast days (ultraviolet light).	See same topic in heat section	Keep clothing on. Apply opaque ointments to exposed skin.	See same topic in heat section.
Windburn A burn-like irritation of the skin.	Cold temperatures and strong winds.	Chapped, reddish skin.	Protect exposed skin.	Prevent further exposure. Apply skin ointments or creams.
Dry skin Skin with its natural oils may be cooled from 32°F down to 20°F before frostbite sets in.	Caused by cold, dry weather.	Dry, chapped skin.	Alleviate by applying a grease or oil based ointment. Excessive washing with soap removes protective skin oils. Use a skin lotion or cream.	Animal or vegetable oils will eliminate dry skin.
Earache Irritation of the eardrum caused by wind.	Wind blowing into ear. More common in cold weather.	Pain in ear.	Protection of the outer ear canal. Wear adequate head protection that covers the ears. Use a plug of cotton or soft tissue.	Prevent further exposure.
Chillblains	Repeated exposure of bare skin at temperatures between the low 60°'s F. and 32°F.	Rosy cheeks. Skin is red, rough, itchy.	Prevent exposure by dressing adequately.	Apply skin lotion or ointment.
Immersion foot (trench foot). Cooling of an extremity at temperatures above freezing causing injury to nerves, muscles, and blood vessels.	Caused by a prolonged exposure to wet conditions, usually temperatures are above freezing. Caused by wearing vapor barrier boots or wearing wet socks and boots for extended periods. The amount of time the extremity is in a wet and immobile condition will determine the extent of injury.	Skin is cold, swollen, waxy in appearance, mottled with burgundy to blue colored splotches. In early stages skin is wet, white, and wrinkled. Walking becomes difficult and painful. Skin becomes numb to touch, deep sensations are lost. Accompanying problems are throbbing pain, burning sensations, cold sensitivity, swelling, muscle weakness, paralysis, atrophy.	Keep feet and hands dry and warm. Change socks, gloves, mittens several times each day. Wipe dry inside of boots when changing socks. Wipe feet dry. Apply foot powder.	Dry feet or hands, and provide warmth. Restore circulation. Bathe to prevent infection, using plain white soap. Dry thoroughly and elevate feet. Do not walk.

Body Upset	Predisposing Factors	Signs and Symptoms	Prevention	Treatment
Hyphydration (dehydration) A major problem in cold environments. Inhibits the body's ability to produce heat—contributes significantly to hypothermia.	Cold environments inhibit thirst sensations. (See same topic in heat section.)	See same topic in heat section.	Adequate fluid intake. (See same topic in heat section.)	Drink water. (See same topic in heat section.)
Hypothermia A lowering of the body core temperature caused by the body losing more heat than it can produce.	Common in wet, windy, cool conditions. Inadequate clothing, or clothing becoming wet. Fatigue or exhaustion. Inadequate food and water intake (hypothermia). Immersion in cold water.	If you feel cold, your body is losing heat. Look for shivering, poor coordination, poor reflexes, stumbling, thickness of speech, careless attitude, irrationality, memory lapses, easily fatigued, lack of enthusiasm, or poor judgement. Blueness of skin, dilation of pupils, decreased heart and respiratory rate, weak or irregular pulse. Watch for a decrease in shivering, followed by rigidity of muscles.	Proper clothing, stay dry, wear a hat, drink warm liquids, seek shelter during storms. Check yourself and others for symptoms. Work slow—do not sweat. Take adequate food, plenty of water and rest.	Remove wet clothing and provide insulation to stop heat loss. Shelter from the elements. Keep body moving. Take warm fluids, if conscious. Breathe warm oxygen. Provide flesh to flesh contact or other external heat source. Gentle evacuation. (No alcohol.)
Frostnip Superficial freezing of skin tissue.	Skin exposed to cold or cold/windy conditions.	A sudden blanching (whitening) of the skin, usually on the ears, fingertips, tips of toes, nose, cheeks, or chin.	Use the 'Buddy System'. Constant personal vigilance for signs. Wear adequate clothing.	Rewarming the affected part by a warm hand, warm stomach, or armpit. Blow warm air on the affected part.
Frostbite of the lungs	Heavy breathing in a cold air environment.	Breathing discomfort, coughing, asthmatic type reactions, coughing up blood.	Pre-warm the air, by breathing through hoods, masks, cloth. Humidify the living environment. Eliminate smoking.	Warm breathing environment.

Body Upset	Predisposing Factors	Signs and Symptoms	Prevention	Treatment
Frostbite Actual skin tissue damage (freezing) by exposure to intense cold. When skin tissue is chilled below freezing. —capillaries constrict, become damaged, circulation is inhibited —ice crystals begin to grow between the individual cells	Skin exposed to cold, or cold/windy conditions. Damage occurs from ice crystals drawing off water from the cells: the cells dehydrate. Cell injury is also caused by the disruption of nutrients, waste elimination, and oxygen intake.	Superficial frostbite—skin is white, waxy in appearance; numbness; resilience to touch. Deep frostbite (freezing)—muscles, bone, tendons, are frozen; affected part becomes rigid, waxy colored, cold to the touch, painless when frozen. If your skins feels cold, frosty, crisp, or resilient, you have got it!	Wear proper clothing. Maintain general body warmth. Shelter against the weather. Good circulation—avoid tight/snug fitting clothing and boots/shoes. Use the buddy system for early detection. Check extremities. Protect face, head, and neck during cold, windy weather. Adequate diet, emphasis on fats and carbohydrates. Avoid perspiration. Keep hands and feet dry. Do not touch metal objects. Wear mittens instead of gloves. Do not smoke or use alcohol.	Identification of injury, protection of injury, evacuation to a medical facility. Minor —rewarm with warm flesh, warm water —stimulate circulation (do not rub) —drink warm liquids (no alcohol) Major —immersion in water, (105°/110°F) until completely thawed —cover with sterile dressing (loose) —do not allow to refreeze
Carbon monoxide poisoning (See same topic in altitude section.)	Use of stoves or internal combustion engines inside of shelters or living environment. Fumes can asphyxiate even when there is adequate oxygen.	Carbon monoxide is odorless, tasteless, colorless, and heavier than air. Often there are no symptoms of this poisoning, but you could feel —burning eyes —dizziness —throbbing temples —sharp headache —pounding pulse —running nose —coughing Look for —"cherry pink" skin appearance —drowsiness —muscle twitching	Maintain good ventilation whenever a stove, heater, or lamp is burning. Never go to sleep with stoves, heaters, or lamps burning.	Get to fresh air. Breathe deeply. Mouth to mouth resuscitation. Warmth and rest.

Wind Chill Chart

ACTUAL THERMOMETER READING °F.

Estimated Wind Speed MPH	50	40	30	20	10	0	-10	-20	-30	-40	-50
	EQUIVALENT TEMPERATURE °F.										
Calm	50	40	30	20	10	0	-10	-20	-30	-40	-50
5	48	37	27	16	6	-5	-15	-26	-36	-47	-57
10	40	28	16	4	-9	-21	-33	-46	-58	-70	-83
15	36	22	9	-5	-18	-36	-45	-58	-72	-85	-99
20	32	18	4	-10	-25	-39	-53	-67	-82	-96	-110
25	30	16	0	-15	-29	-44	-59	-74	-88	-104	-118
30	28	13	-2	-18	-33	-48	-63	-79	-94	-109	-125
35	27	11	-4	-20	-35	-49	-67	-83	-98	-113	-129
40	26	10	-6	-21	-37	-53	-69	-85	-100	-116	-132

Wind speeds greater than 40 MPH have little additional effect

LITTLE DANGER FOR PROPERLY CLOTHED PERSON	INCREASING DANGER	GREAT DANGER
	DANGER FROM FREEZING OF EXPOSED FLESH	

To use the chart, find the estimated or actual wind speed in the left-hand column and the actual temperature in degrees F. in the top row. The equivalent temperature is found where these two intersect. For example, with a wind speed of 10 mph and a temperature of -10°F, the equivalent temperature is -33°F. This lies within the zone of increasing danger of frostbite, and protective measures should be taken. It is emphasized that the wind-chill chart is of value in predicting frostbite only to exposed flesh. Outdoorsmen can easily be caught out in 30° temperature. Winds of 30 mph will produce an equivalent wind-chill temperature of 2° below zero.

Wind Comparison Chart

WIND SPEED ESTIMATION	M.P.H.
Smoke rises vertically	0-1
Smoke shows wind direction	1-3
Wind felt on face, grass or leaves rustle	4-7
Leaves in constant motion, light flag extended by breeze	8-12
Dust or snow or leaves are raised	13-18
Small trees in leaf sway; crested wavelets on inland waters	19-24
Large branches in motion; white caps on waves; tents strain	25-31
Whole trees in motion; snow rises in air	32-38
Twigs break off trees; walking generally difficult	39-46
Branches break off trees; high waves and tides	47-54
Trees uprooted	55-63

Body Upset	Predisposing Factors	Signs and Symptoms	Prevention	Treatment
Hypohydration (dehydration) A lessening of the body's water table. Inhibits normal body functions.	The body loses water rapidly in hot weather through perspiration, respiration, urination, bowel movement.	First stage, mild, 1-5% loss of body water —thirst, loss of appetite, weakness, impatience, nausea, irritability, drowsiness, loss of efficiency, general discomfort, yellow urine. Second stage, moderate, 6-10% loss of body water —dizziness, headache, dry mouth, dark yellow urine, absence of salivation, shortness of breath, impaired speech, inability to walk, tingling of limbs, blueness of tongue and lips, mental impairment. Third stage, severe, 11-20% loss of body water —delirium, tongue swollen, spasms, deafness, dimming of vision—**death**.	Adequate (continuous) water intake. Proper diet (replenish salt loses). Slow down, rest in shade. Keep clothes on. Breathe through nose. Ration your sweat, not water.	Drink water. Replace body salts. Find shade.
General Signs and Symptoms of Body Heat Strain When it's hot, be on the lookout for the following signs in yourself and others —general fatigue and weariness (all muscles and joints) —headache, nausea, faintness —personality changes (irritability, carelessness, immature outbursts) —abnormal pulse (weak, rapid, irregular) —change in appearance (pale skin, flush, bluish lips or fingernails) These signs may or may not be caused by heat strain. Keep a margin of safety and treat them as such.				
Heat syncope (fainting) Injuries can result from falling.	Prolonged standing or exercising in hot environments. The blood pools in the limbs, reducing flow of blood to the brain.	Feeling faint, giddiness, fatigue.	At first sign of dizziness, lie down with feet slightly elevated. Take plenty of fluids and water, follow proper diet, and get adequate rest in the shade.	Lie down or rest in the shade. Drink water.
Heat weakness (strain) Results from not being acclimatized to heat.	Occurs in very hot, humid weather.	Easily fatigued, headache, mental and physical inefficiency, insomnia, poor appetite, heavy sweating, high pulse rate, general physical weakness.	Allow time to acclimatize. Drink plenty of fluids and water. Follow a proper diet and get adequate rest.	Cooler environment. Drink plenty of water. Replenish body salt loss. Lie down—rest.
Sunburn A first or second degree burn will retard or prevent sweating.	Exposure to sunlight, excessive ultraviolet radiation. Occurs even on cloudy days. A strong wind will make the burn more severe.	Burns, broken skin, infection-prone blisters.	Keep clothes on. Wear lightweight and light colored clothing. Protect head and neck. (Sunburn occurs on cloudy days as well.) Apply opaque ointments or lotions containing aminobenzoic acid (PABA) to exposed skin.	Cool the skin. Apply approved burn medications. Avoid further exposure.

Body Upset	Predisposing Factors	Signs and Symptoms	Prevention	Treatment
Sunblindness Burning of the mucouse membrane lining the surface of the eyelids and covering the front part of the eyeball.	Eyes exposed to too much solar radiation or reflected sunlight.	Redness, burning, watery or sandy feeling eyes. Headaches, poor vision, pain and swelling, abnormal sensitivity to light, smarting-itchy eyelids.	Wear proper sunglasses, with side protection. Or, improvise glare shades by cutting small eye slits in any material suitable to tie around the head.	Cover the eyes. Keep in dark environment. Cold compresses. Boric acid ointment. Antibodies to prevent infection.
Heat cramps Chemical unbalance, body salts deficiency is the primary cause.	Often due to strenuous activity in high heat and high humidity when heavy sweating depletes salt level in the blood and tissues. Also due to long term fluid and salt deficiency.	Heavy sweating. Muscle cramps in legs and abdomen. Painful, voluntary muscle spasms. Upset stomach. Skin is cold and clammy.	Adequate liquid intake. Proper diet. Replacement of body salt loses.	Replenish body salts. Plenty of water and food. Shade and rest. Gentle massage.
Heat exhaustion Circulatory unbalance. Disruption of circulatory system, causing insufficient supply of blood.	Too much exertion in hot environments. Usually a cumulative building process. Other heat upsets occur first.	Weakness, fatigue, dizziness, nausea, cold-clammy skin, possible cramps, delirium, heavy sweating, face flushed then pale, possible vomiting, loss of appetite, lightheadedness, rapid, weak, or irregular pulse.	Minimize activity—find shade and rest. Take plenty of water. Follow a proper diet—replenish body salts loss, high energy snacks.	Find shade and rest. Lie flat on back — elevate feet. Keep warm. Take plenty of cool water. Salty water as tolerated (½ teaspoon of salt/quart of water).
Heat stroke Nervous system unbalance. Body temperatures rises rapidly, pulse pounds, blood pressure is high.	Usually caused by excessive sweating under conditions of high heat. high humidity, direct sunlight, and muscle activity. **Heat Exhaustion** 1. Pale face 3. Moist cool skin, sweating profusely 2. Weak pulse 4. Low temperature **Care:** 1. Position head level or low 3. Keep warm 2. Give salt in solution 4. Call doctor	Sweating stops and skin becomes dry. Weakness, nausea, headache, pounding pulse, high temperature. Skin flushed, then turns ashen. Delirium or coma, unconsciousness. **Sunstroke** 1. Red face 3. Hot, dry skin 2. Rapid strong pulse 4. High temperature **Care:** 1. Elevate head 2. Do not give stimulants 3. Cool body by cold bath or cold applications 4. Call doctor	Do not allow body temperature to rise. Slow down and get plenty of rest in the shade. Take plenty of water and follow proper diet—replenish body salts loss.	This is a medical emergency. Cool the body anyway possible. Wet clothing and fan to increase evaporative cooling. Find shade, massage extremities. Hydrate with salty fluid (½ teaspoon of salt/quart of water). Do not take alcohol or aspirin. Evacuate and get to medical facility as soon as possible.

Body Upset	Predisposing Factors	Signs and Symptoms	Prevention	Treatment
Hypoxia Lack of sufficient oxygen in the body cells or tissues. The function of various organs, including the brain, is impaired. Types of hypoxia —hypoxic (altitude) —hypemic (anemic) —stagnant —histotoxic	Flying above 12,000 feet in an un-pressurized aircraft without supplemental oxygen. Landing or living at altitude. Failure of the body tissues to use oxygen efficiently because of cellular respiration impairment. (Alcohol, drugs are often directly responsible.) (Histotoxic.) Insufficient partial pressure of oxygen in the inspired air (hypoxic). Reduction of red blood cells, caused by carbon monoxide poisoning or excessive smoking (anemic). Failure of the circulatory system to pump blood and oxygen to the tissues, caused by shock, long periods of positive pressure breathing, or excessive G-forces (stagnant). Pathological conditions interfering with normal ventilation of the lungs, such as strangulation, drowning, pneumonia.	Increased breathing rate, headache, fatigue, light-headed or dizzy sensations. Listlessness, tingling or warm sensations. sweating, poor coordination, impairment of judgement, behavior changes. Feeling of well being (euphoria). Loss of vision, or reduced vision, sleepiness, blue discoloration at the fingernail beds and lips (cyanosis). "Cherry Pink" discoloration is a sign of carbon monoxide poisoning.	Use of supplemental oxygen above 12,000 feet. Adequate supply and proper functioning of on-board oxygen systems. Abstinence from alcohol prior to flight. Abstinence from smoking during and prior to flight. Only use medications (drugs) prescribed by a flight surgeon or an aviation medical examiner. Check for leaks in heater and exhaust manifold prior to flight.	Emergency descent to altitudes below 10,000 feet. Carbon monoxide poisoning —use supplemental oxygen —descent —ventilate cabin —get medical help —get aircraft checked
Hyperventilation (over breathing) Excessive breathing is a disturbance of respiration that may occur as a result of emotional tension or anxiety. Breathing too rapidly or too deeply causes excessive elimination of carbon dioxide.	Lung ventilation increases, due to emotional stress, fright, or pain, reducing the normal balance of carbon dioxide in the blood. Also occurs as a result of the body's normal compensatory response to hypoxia. Flying under stress or at high altitude.	Rapid, deep breathing, dizziness, hot/cold sensations. Tingling of the hands, legs, and feet. Tetany, nausea, sleepiness, blurring of vision. Muscle spasm, rapid heat rate, unconsciousness.	Normal breathing. Know and believe that over breathing can cause hyperventilation problems.	Consciously slow breathing rate until symptoms clear, then resume breathing at a normal rate. Breathing can be slowed by breathing into a bag, or by talking aloud.

Body Upset	Predisposing Factors	Signs and Symptoms	Prevention	Treatment
Carbon monoxide poisoning (see hypoxia). Also see same topic in cold section.	Carbon monoxide is produced by —defective mufflers —faulty exhaust systems —heater leaks (See hypoxia).	The most common source of carbon monoxide intoxication in an aircraft is tobacco smoke. Tobacco lowers a pilot's tolerance to altitude. Tobacco also lowers sensitivity of the eyes and reduces night vision by 20%. It increases the body's demand for oxygen and reduces the body's oxygen supplies (see hypoxia).	(See hypoxia) Do not smoke. If you are a smoker, use oxygen above 5,000 feet.	(See hypoxia) Supplemental oxygen. Descent to lower altitude. Ventilate cabin. Get medical help. Have aircraft checked.

Disorientation (vertigo) A state of temporary spacial confusion resulting from misleading information sent to the brain by various sensory organs.	Flight factors contributing to sensory illusions
	—acceleration/deceleration changes
	—cloud layers
	—low level flight over water
	—frequent transfer from IFR to VFR
	—unperceived changes in flight attitude and/or altitude
	Factors contributing to visual illusions
	—windshield optical characteristics
	—rain on windshields
	—effects of fog, haze, dust on depth perception
	—angle of glide slope: steep angle-apron appears farther away and runway longer. Shallow angle-apron appears nearer and runway shorter.
	—width of runway-wide runways give nearer to ground feeling
	—variations in runway lights
	—runway slope:
	  upslope—tendency to land short.
	  downslope—tend to land long.
	—terrain slope:
	  upslope—tend to land long.
	  downslope—tend to land short.
	—landing over water—tendency to land short.
	—autokinetic phenomenon—apparent motion of fixed lights in the dark.
		You are susceptible to confusing disorienting experiences when
		—you are seated on an unstable moving platform at altitude and your vision is cut off from the earth, horizon, or other fixed reference.
		—you are exposed to certain angular accelerating or centrifugal forces which you cannot distinguish from gravity.
		Disorientation occurs most often in instrument conditions created by rain, fog, clouds, smoke, or dark nights. It is aggravated by lack of instrument experience or training, unfamiliarity with the aircraft or flight situation, fear or worry, and excessive head movements.
			Understand the nature of vertigo and the causes.
			Avoid flight conditions which tend to cause vertigo.
			Obtain instrument flight instruction—maintain proficiency.
			Have faith in your instruments.
			Do not trust your sensory organs.
			Remember, it can happen to anyone.
			Avoid sudden head movements, especially when changing altitude.
			Use either VFR or IFR—do not oscillate from one to the other.
				Read your instruments.
				Put your faith in your instruments and not in your senses.

Body Upset	Predisposing Factors	Signs and Symptoms	Prevention	Treatment
Decompression sickness (dysbarisms) Caused by gas expanding or evolving within the body during ascent and descent. Basically, it is the inability of the body to equalize with pressure changes causing abdominal pain, toothache, pain in the ears and sinuses.	Decompression sickness can occur in one of two ways • trapped gas—during ascent and descent free gas expands or contracts in body cavities • evolved gas—caused by gases, mainly nitrogen, some oxygen, carbon dioxide, and water vapor, escaping from solution in the blood and other body fatty tissues, at high altitudes (above 18,000 feet) causing —bends (bubble formation around joints or muscles) —chokes (bubble formation in lung tissue) —paresthesias (skin sensations) —central nervous system problems	Headache, ear pain, feeling of abdominal fullness. Overweight people are more susceptible to evolved gas decompression sickness. Severe pain. Burning sensation or stabbing pain in the chest area, coughing, difficulty in breathing. Tingling, itching, cold and warm sensations, red rash. Lines or spots before eyes, field of vision blurring, dull and persistent headaches.	Slow down rate of ascent. Do not eat too quickly before a flight. Do not eat too much (swallowed air increases with each bite). Avoid large quantities of fluids, especially carbonated drinks, beer. Do not eat gas-forming foods: beans, cabbage, onions, raw apples, cucumbers, melons, or any greasy foods. Avoid chewing gum while ascending (could result in swallowing a great deal of air). If you have a cold and fly, you can expect ear problems. Stay on the ground. If you must fly, then fly at lower altitudes.	Descent to lower altitude. Remain quiet. Keep the affected area immobile. Toothache? See a dentist. Call the USAF School of Aerospace Medicine (24 hour hot-line 512-536-3278) if you or any passenger experience evolved gas type decompression sickness. To clear ears, swallow, yawn, or tense muscles of the throat at intervals. Or, close mouth, hold nose, and blow.
Motion sickness Motion sickness causes the sense of balance to be lost. Flying efficiency is jeopardized.	Caused by the continued stimulation of the inner ear, particularly in turbulent weather. Can be a result of anxiety.	Disoriented, nauseated, vomiting, headache. Perspiring freely. Saliva collects in mouth. Loss of desire for food.	Overcome by experience. **Caution**, if you are the pilot and are susceptible to motion sickness, **do not** take preventive medications. They may make you drowsy or depress brain functions. Motion sickness drugs can cause temporary deterioration of judgement.	Open air vents, loosen clothing, use oxygen. Keep eyes on a point outside the airplane. Avoid unnecessary head movements. Cancel flight plan—land as soon as possible.
Drugs Drugs will distort mental processes and will cause pilot error.	Even legitimate medications can jeopardize safe flight. Eating certain foods in combination with certain medicines produces dangerous effects.	Drowsiness, alertness reduced. A feeling of high spirits and false confidence, while actually crippling judgement. Reactions slow down or speed up. Nausea, vertigo.	The need for medicines implies the presence of an illness. If you are ill, you should not be flying.	Take no medicine or drugs before flying without consulting a flight surgeon.

Body Upset	Predisposing Factors	Signs and Symptoms	Prevention	Treatment
Alcohol Alcohol hampers the ability to think, reason, and perform tasks. It is the cause of 20-40% of air crashes, and detracts from the ability to make quick, successive decisions.	Depresses the central nervous system; acts upon the body like a general anesthetic. Interferes with the brain's ability to utilize oxygen (see hypoxia). Alcohol problems are compounded if you are fatigued, hungry, or under stress.	Thinking becomes sluggish. Coordination problems.	Do not drink and fly. Remember the following facts —bloodstream will absorb 80-90% of alcohol consumed within 30 minutes —the body requires 3 hours to rid itself of the alcohol contained in one drink or one beer —body functions require up to 2 days to recover from heavy drinking	Do not drink alcohol in any form during the 8 hour period before flight. Do not overindulge during the 24 hours before flight.

Blood-Alcohol Chart

Number of Drinks

Weight (lb.)	1	2	3	4	5	6	7	8	9	10
100	0.043	0.087	0.130	0.174	0.217	0.261	0.304	0.348	0.391	0.435
125	0.034	0.069	0.103	0.139	0.174	0.209	0.242	0.278	0.312	0.346
150	0.029	0.058	0.087	0.116	0.145	0.174	0.203	0.232	0.261	0.290
175	0.025	0.050	0.075	0.100	0.125	0.150	0.175	0.200	0.225	0.250
200	0.022	0.043	0.065	0.087	0.108	0.130	0.152	0.174	0.195	0.217
225	0.019	0.039	0.058	0.078	0.097	0.117	0.136	0.156	0.175	0.195
250	0.017	0.035	0.052	0.070	0.087	0.105	0.122	0.139	0.156	0.173

Blood-alcohol chart shows estimated percentage of alcohol in the blood by number of drinks in relation to body weight. 1 drink equals 1½ oz. of liquor, 12-oz. bottle of beer or 3 oz. of wine. Count 1 drink of over-proof rum as 2 drinks.

Impairment is reached before 0.080%
For Safety: Don't Drink and Fly

Alcohol Burn-off

Hours since first drink	1	2	3	4	5	6
Subtract from blood-alcohol	0.015	0.030	0.045	0.060	0.075	0.090

Examples:

Wt.	Drinks	Time	
150 lb	4	3 hr	= 0.116−0.045 = 0.071%
150 lb	6	4 hr	= 0.174−0.060 = 0.114%

1. Alcohol will affect your flying ability. The percentage of alcohol in a person's blood is a guide to how much your flying may be impaired.
2. To estimate the percentage of alcohol in the blood follow these directions:
   a. Count your drinks (1 drink equals 1½ oz. of liquor, 3 oz. of wine or 12 oz. of beer;
   b. The blood-alcohol chart shows the number of drinks in relation to your body weight and gives the percentage of blood-alcohol in your body;
   c. Subtract from this the percentage of alcohol burned up in your body during the time elapsed since your first drink. This would give you the percentage of blood-alcohol. Critical point is reached at 0.08 per cent.

Body Upset	Predisposing Factors	Signs and Symptoms	Prevention	Treatment
Fatigue A depletion of body energy reserves, leading to lower efficiency and performance.	Caused by —mild hypoxia —physical stress —psychological stress —depletion of physical energy	Acute fatigue (short term) —tiredness after a period of strenuous effort, excitement, or lack of sleep. Chronic fatigue (long term) —continuous strain —weakness —tiredness —palpitations of the heart —breathlessness —headaches —irritability —can create stomach or intestinal problems —aches and pain throughout the body	Proper, well balanced diet Adequate rest and sleep. Good physical condition. Positive mental attitude.	Acute fatigue —rest —do not fly Chronic fatigue —**do not fly** —see a physician
Vision The brain can create illusions by misinterpreting images.	Vision can be seriously impaired by —fatigue —colds —vitamin deficiency —alcohol —stimulants —smoking —medication	Recognize the predisposing factors that can cause impairment of vision.	Recognize and believe that your brain and eyes can play tricks on you.	Trust your instruments.
Night flight Night flying requires different visual techniques than day flying.	Autokinesis, one of the special visual night flying hazards, resembles vertigo. It occurs when you stare at a light in a dark sky. After a while, you will get a feeling that either you or the light is moving.	Know and recognize that vision upsets can occur at night.	Autokinesis —keep your eyes moving —do not stare at a single light too long —eyes will need about 30 minutes to adjust back to maximum efficiency after exposure to bright light —avoid looking directly at bright lights—close one eye when exposed to bright light —lightning flashes affect night vision—when near storm clouds, turn up cockpit lights —remove sunglasses after sunset	Trust your instruments. Also, after some practice, you will find that you can see things more clearly and definitely at night by looking slightly to one side of them rather than straight at them.

Body Upset	Predisposing Factors	Signs and Symptoms	Prevention	Treatment
Noise Unless ear protection is used, hearing loss will occur over a period of time. If you fly more than 5-8 hours per week, without ear protection, you can expect to develop hearing loss problems within 10 years.	Primary sources of noise in light aircraft —squeaks and rattles —exhaust —propellers —ventilation system —air turbulence around fuselage	Deterioration of hearing can be detected best by special testing with an audiometer in the frequency ranges which are above the pitch of the human voice.	Regular hearing check-ups. Use ear defenders (such as ear plugs or ear muffs). Protect yourself against any noise which produces pain in the ears. Avoid unnecessary exposure to noise. Lower the volume on your earphones or speaker when possible, especially tone signals of navigational aids and heavy static.	If hearing problems or deficiences develop, see a physician.

CHAPTER V
The Mental Aspects of Emergency Response and Survival

There is really no way to accurately predict any person's behavior or thought processes during sudden mental and physical stresses. Reactions to every situation will vary greatly with every circumstance and every individual. In general, there are two aspects of emergency response which should be considered by every pilot. They are: (1) mental preparation to every day flight; and (2) response to an emergency situation.

First, mental preparation is as important to safe flight as the condition of the aircraft. Anger, fear, frustration, depression, worry, and anxiety all affect a person's ability to concentrate. Anyone who brings their problems into the cockpit is easily distracted and less able to adjust to any kind of stress. The relationship between mental functions and physical capability has already been discussed. A good example of this might be a situation where an individual's eyes may not interpret what they are actually seeing on the instrument panel. Current research even points out that emotional disturbances can hamper a person's ability to adjust to altitude.

Another very important point in this regard pertains to the passengers. It is not uncommon for normally calm individuals to be completely unnerved during relatively routine procedures, especially if they do not frequently fly. In the case of several passengers, one of them can quickly affect the others with their reactions. Those who fly regularly should be aware that others are not as confident or comfortable in the air as they themselves are and certain precautions should be taken to minimize discomfort and worry. Avoiding sudden control movements, uncertainty in procedures, and any behavior that might undermine passenger confidence are worthy of consideration. This is especially pertinent if the pilot expects passengers to constructively react to any emergency situation.

Since there is really no way to set up a real test situation and effectively monitor it, investigators must rely heavily upon details recounted by survivors of aircraft emergencies. Unfortunately, the human mind is influenced by a great many things, and more often than not when recounting our own experiences, accuracy and the importance of detail are distorted.

For these reasons, accurate predictions on how any one person may react in a life threatening situation are improbable. What is known is that under repeated stress situations, people will make more and more effective adjustments. First reactions are the most important, however, and will be discussed here briefly.

There are many things present in everyday life to provide indications of stress performance. We must learn to read these indicators and make necessary adjustments.

The Mechanics of Mental and Physical Response to Life Threatening Situations

One of the few areas of knowledge that survival instructors universally agree upon is the need for controlling the mind in an emergency environment. This is virtually accepted by every advocate of emergency preparedness and survival training. The following is an outline of facts and helpful information which will enable any pilot to better understand how to effectively deal with any emergency situation. Although there are four stages in the Initial Response Phase, the remaining phases may be far more complex as a result of the initial action.

Initial Response Phase

Stage I—Alarm: a state of alertness as a result of some stimulus.
- Visual
    - decrease in oil pressure or manifold pressure
    - deterioration of weather conditions
    - smoke in the cockpit
- Sound
    - rough running engine
    - loud noise
    - silence

> *Anxiety appears as a natural reaction to what could happen.*

Stage II—Reaction: the physical body gears up for action. The entire body, both physical and mental feels that there is a possibility that something is going to happen.
- Muscles tighten
- Sweat glands close down
- Sugar is released for energy
- Adrenelin starts to flow
- Heart rate increases

> *If allowed to progress, anxiety will turn to overt fear.*

NOTE: This is the point at which training and or experience plays a major role in subsequent events. If a pilot has established standard emergency procedures and a sequential method of attacking the problem, then the response is natural and positive. Checklists and sequential procedures are standard in all military flying. The reasons are obvious. The response must be instant, natural and must not consist of any involved thought process or decision making.

> *There is no substitute for practice or experience.*

Symptoms of Fear

Physical Symptoms
- *Quickening of pulse, breathlessness*
- *Dialation of pupils*
- *Increased muscular tension and fatigue*
- *Perspiration of hands, feet and armpits*
- *Frequent urination*
- *Dry mouth and throat, high pitched voice with stammering*
- *Empty stomach causes "butterflys" and faintness*
- *Nausea—vomiting*

Mental Symptoms
- *Irritability, increased hostility*
- *Talkative at early stages, later speechless*
- *Laugh and cry hysterically*
- *Confusion, forgetfulness, inability to concentrate*
- *Feelings of unreality, flight, panic and sometimes stupor.*

> *"No passion so effectively robs the mind of all its powers of acting and reasoning as fear."*
> —Edmund Burke.

Stage III—Response and Options: the fight or flight syndrome—adrenelin is released into the system.
- Methodical approach to the problem with precision moves and sequential procedures.
- Scatterbrained thinking with no definite plan and refusal to believe the situation is really that bad or is happening at all.

- Complete panic with frozen limbs and mind. (Weakness, crying, trembling, nausea, vomiting.) These can be compared to the ground environment where an individual might turn and run, stand and do something positive or merely turn into a shaking blob. It must be understood that anxiety and fear are perfectly natural feelings that can and must be controlled.

Stage IV—Rest: sharp emotional letdown after high energy output.
- Eventually, this will come whether wanted or not.
- In many cases, this will be a complete drain, both physically as well as emotionally.
- Chances are shock will occur. Be prepared for it, and treat it before it occurs.

Each Person Has an Individualized Comfort Zone

Humans are basically creatures of habit, dependent on routine, organization, and some degree of discipline. Once these are established, they become comfortable with their behavior and surroundings. People generally become skilled and proficient only after a period of slow, deliberate actions while learning what is expected. Everyone establishes these "comfort zones" both physically and mentally. It follows then that whenever a person is placed in a situation that forces use of the outer limits of his/her comfort zone, anxiety and stress are created.

When drastic changes occur during emergency situations and initial response is automatic, a brief time period is gained for composure and some type of positive thought process is then possible. The most successful reactions to emergency situations have been those that recounted slow deliberate actions that brought about results.

Comfort zone limits vary with each individual in accordance with experience and knowledge. Everyone is unique in abilities and since the human being is basically a creature of habit, most people will rarely go outside of their comfort zones. Herein lies the difficulty of functioning in any emergency environment. Very often, every decision is important and there is little room for mistakes.

Are You a Comfort Zone Expander or Are You Content with Your Everyday World?

At best, decision making is always difficult and even more so in times of emotional upset. The best way to prepare mentally is through self-discipline and practice during everyday routine. Look for opportunities to expand your comfort zone into areas that you may be unsure of now.

Are you a person who continually takes the easy way out, or do you tackle everyday problems with a positive attitude and direct approach? These questions may even be applied to personal relationships and situations that develop there. The answers may give you a good indication of your ability to react under stress.

> **Good habits in training and everyday life pay off in an emergency.**

Controlling Fear

As any individual approaches comfort zone limits, chances are that individual will experience an unconscious level of fear known as anxiety. This level is characterized by feelings of uneasiness, general discomfort, worry and/or depression. This can vary in intensity, duration and recurrence. If this anxiety is allowed to progress, it may become an overt fear. Fear, unchecked and reinforced by other thoughts or stimuli, can quickly turn to panic with complete loss of reasoning. Fearfulness turned to blind panic can cause an experienced person to injure or kill oneself, perhaps others, in the intensity of terror.

Training, including knowledge and experience gained in simulated emergencies (practice 180° turns at altitude or power off landings), reduces the unknown and helps to control fear. When fantasy distorts a moderate danger into a major catastrophe, behavior can become abnormal. If this happens, people may be reacting to feelings and imagination rather than the problem. In such cases most people tend to underestimate (rather then overestimate) danger which readily leads to reckless, even foolhardy behavior.

There are many ways to fight fear, but one of the best is to meet it head on. **Recognize fear as a natural phenomenon.** Try to establish just why it is that you are afraid and accept it as the defense mechanism that it is. Some professionals in the behavior field believe it is beneficial to get mad at your own fears. At least it is a positive action that puts the fear into proper prospective.

Ask yourself if others have conquered the same situation. There are no clear cut lines between recklessness and bravery, caution and panic. **That is why it is essential to maintain proper mental control.**

> *To return from an emergency situation you need the following:*
> - *Knowledge of the entire situation*
> - *Tools and/or equipment for emergency response*
> - *A positive mental attitude (a will to live)*
>
> *Personality Traits of a Survivor*
> - *Can make decisions*
> - *Can improvise*
> - *Can adapt and "make the best" of the situation*
> - *Has patience—can keep cool, calm, collected*
> - *Is prepared—hopes for the best, but prepares for the worst*
> - *Knows own special fears and worries—and most importantly—can control them*

Ways to Control Fear in Self

- Do not try to physically or mentally run away from the situation. Recognize fear for what it is and accept it. Try to learn what your reactions are likely to be by looking at your daily habits.
- Learn how to make decisions quickly and logically by establishing good habits. Take positive action to take control of the situation instead of letting it control you.
- Develop self-confidence by continually expanding your comfort zone to encompass experiences you are not familiar with. Define your fears and recognize them.
- Realize "it can happen to me" and be prepared. Be properly equipped and prepared at all times. Always have a number of options in your plan. Prepare for the worst and hope for the best.
- Keep informed and increase your knowledge to reduce the unknown.
- Have procedures mapped out so that you will be busy. If not physically, then mentally.
- Set realistic goals.
- Realize that teamwork always accomplishes more than a one person show.
- Use affirmative self talk. Talk positively about your actions.
- Do not be afraid of spiritual faith.

- Gather as much information about your situation as you can.
- Cultivate good survival oriented attitudes. The main goal is survival with everything kept in perspective. Discomforts of the moment are only temporary.

Ways to Control Fear in Others
- Cultivate mutual support.
- Use good leadership practices.
- Practice discipline.
- Use positive contagion to advantage.
- Do not indicate resentment of others' reactions. Accept a person's right to individual feelings.
- Do not chide others. Accept a person's limitations.
- Comfort them without encouraging them to feel sorry for themselves.
- Involve them in simple tasks.

Emergency Management Requires
Four Mental Steps
S — Stop
T — Think
O — Observe
P — Plan

Each Person Acts in Accordance with Individual Beliefs

Many adult fears and related misconceptions are based upon stories and statements originating during early years of life and from unknowledgeable sources. These thoughts, statements, and stories are stored in the subconscious. This is why survival and emergency preparedness training is so important.

Negative feedback inadvertently given during childhood or during any learning process may have a profound effect on behavior and reaction. Negative reaction to fledgling attempts at independence and self-reliance are good examples of programming future adults for poor performance. The same holds true after a person reaches maturity and receives negative feedback for some type of performance. Negative self talk has a dramatic negative effect on performance. All people must learn to program themselves and others more positively, remembering that individuals act or react in accordance with what each believes to be true.

The Subconscious Mind Mirrors Self-image

The mind is constantly at work picking up information and skills. The subconscious will never erase whatever is intentionally or unintentionally placed there. In any situation, actions are a reflection of how an individual perceives the environment and the individual in it. If the self-image is strong, the individual will be strong. If the picture is of a weakling, the individual will be weak. More importantly, individuals who think they can not survive probably will not even try.

Every Survivor Sets a Goal

Hundreds of accounts and personal interviews with those who have undergone trying experiences, mental and physical, reveal that nearly every survivor has set some type of goal. The goal may have been just to live or it may have been a combination of staying alive and accomplishing something else. Philosophically, life is made up of setting one goal after another and taking the trip in between. Many of the POWs released after the Vietnam conflict related numerous self-set goals which gave meaning and purpose to staying alive.

The goal setting process is vital and has been used extensively, but there are still many who cheat themselves on its full effect. It has been said that "life is truly not the setting of personal goals, but the journeys to reach those goals that makes living worthwhile." If this is true, then constantly learning to re-adjust goals provides incentive, not only to live, but to give purpose to existence. Likewise, a survivor must re-adjust goals to suit the situation. Many people have survived tremendously difficult ordeals after crashing only to die just after being rescued. Their main goal could have been to be rescued. Perhaps they failed to look beyond that goal and establish others. They simply let down and gave up too soon.

We Choose Our Path

Each of us as an individual, has the ability to evaluate performance potential during a time of emergency or stress. We are creatures of habit, and do things regularly that will give good indications of how we will react to emergencies.

In nearly all cases, we choose the path we follow after the onset of an emergency. All of us have the ability to use our mind as a valuable tool or have it work against us as a feared enemy. It is never too late for changing our habits, values, and attitudes that will produce positive mental actions during emergencies.

> *Maintaining a Positive Mental Attitude (PMA) is the key to successful emergency response and better everyday living.*

Positive Mental Attitude

• Goal •

Keeping mind and body at peak efficiency

Mental

I choose to have self-confidence.
I choose my positive self-image.
I establish my comfort zones, constantly seeking to expand them.
I maintain positive habits, values and attitudes.
I set realistic goals.
I make positive decisions and choices.

Physical

I will provide adequate air and control my breathing.
I will maintain my body temperature with adequate shelter.
I will provide adequate mental and physical rest to conserve energy.
I will conserve body water levels thru my activities.
I can exist without food for longer than I realize.

All Must Be Kept in Proper Balance and in Perspective

Notes

CHAPTER VI
Emergency Preparedness Skills

Survival First Aid

Survival first aid is needed when people are unable to get professional medical help for a period of time. Individuals must be prepared to help themselves and others.

Every survival situation has first aid requirements. Minor injuries or ailments become major problems under survival conditions, and prompt attention becomes more critical than under ordinary conditions.

Treatment performed under survival conditions may be substandard compared with modern medical standards, but it is the best treatment available and should increase survival expectancy.

Control of Bleeding

Control of bleeding is always important — even more so under survival conditions because transfusions are usually not possible.

The flow of blood should be stopped immediately, using a method effective for the type and amount of bleeding. Simple pressure over the wound, with or without a dressing, often works, especially for venous or capillary bleeding. Minor arterial bleeding can be controlled with local pressure and extremity wounds may be elevated above the heart and pressure applied. More serious arterial bleeding can be temporarily controlled by pressing the artery against a bone. Direct pressure stops 95 percent of all bleeding.

Wounds

Clean wounds promptly using clean, preferably sterile, water. Irrigating the wound rather than scrubbing hard will minimize additional tissue damage. Wash foreign material from wound to remove sources of infection and wash skin around wounds thoroughly before bandaging.

Pain, Injury, and Shock

Pain is warning of injury or damage to some part of the body. Pain can be controlled, and if the survival problem is grave enough, it is possible to override pain and carry on.

Reduce pain by
- understanding its source and type.

- recognizing that it is a discomfort that must be tolerated temporarily.
- concentrating on what must be done — think, plan, keep busy.

Rest the injured part of the body in the most comfortable position. Splints or bandages may be needed for immobilization. Warmth or cold can ease pain. Apply warm and cold water to the painful area to determine which is most comfortable.

Some degree of shock is present in all injuries. Shock may be the most serious consequence of injuries. It is the body's reaction to injury or emotional disturbance. Be familiar with signs and symptoms of shock in order to recognize and deal with it. The best approach is to treat all injured patients for shock. It won't hurt, and it may help.

In shock, pulse is weak and rapid. Breathing is shallow, rapid, and irregular. Skin is pale and feels cold and moist. "Cold sweat" is common. The eyes appear vacant and may be dilated. The victim feels weak, faint, or dizzy, and may be restless, frightened, and anxious. Thirst and nausea often accompany shock. As shock deepens, the victim may become quieter and slip into unconsciousness.

Do not wait for these symptoms before treating for shock. Expect it in every serious injury and treat simultaneously. The patient should be as comfortable as possible with head lowered to insure adequate blood supply to the brain (unless there is a head injury). Victim should be kept warm, but not hot, and protected from weather.

In emergencies, shock often occurs immediately. This shock is mental and can occur without physical injury. Post-impact shock varies widely in severity, but is almost universal.

Injuries to Bones and Joints

Under survival conditions, initial immobilization of fractures, dislocations, and sprains is part of the ultimate treatment, rather than a temporary measure, to prevent further injury en route to the hospital. Proper immobilization hastens healing and improves the chances of the injured part being used again. Under survival conditions, the injured part must be immobilized for relatively long periods of time, yet still permit patient movement.

Traditional first aid procedures call for immediate splinting and immobilization even if the limb is disfigured. In survival, do everything to slowly and surely return broken bones to original physiological position. If this is not done, the limb may become permanently disfigured or add to medical complications. Protruding bones which have broken through skin must be surgically replaced.

Burns

Burns are serious injuries and are common during emergencies. They are very painful and susceptible to infection. They cause large losses of body fluids and salts. Treat to relieve pain and prevent infection by covering burn with clean dressing of any type which permits movement and vital survival functions.

Maintaining body fluid and salt levels is essential in burn treatment. During emergency, fluids must be administered by mouth. Patients should be given adequate water immediately after treatment and blisters that develop should not be broken as they provide a sterile, natural covering.

Hygiene

Good hygiene and cleanliness are essential to survival. They safeguard good health by minimizing internal and external infection. The smallest abrasion of an unclean body can eaily become infected and present a major survival obstacle. Keep hands as clean as possible, clean under fingernails, and sponge face, armpits, crotch, and feet at least once a day.

Protection Against Intestinal Ailments

Diarrhea, food poisoning, and other intestinal diseases are common and often serious. They are caused by contaminated food, water, or other beverages.

To guard against intestinal diseases
- keep body and hands clean. Avoid overhandling of food. Keep fingers out of the mouth.
- use water treatment tablets or boil water.
- wash and peel all fruit.
- do not keep food for long periods after preparation.
- boil eating utensils in water.
- keep living area clean. Keep flies and other vermin off food.
- dispose of human waste and garbage carefully.
- mix powdered charcoal with a drink to stop diarrhea.
- brush teeth regularly. Soap or table salt and soda substitute for toothpaste, and a small, green twig chewed to pulp at one end makes a good toothbrush. Rinse mouth after eating.

Parasitic Disease

Disease may be one's worst enemy in an emergency. It is necessary to know what diseases are present in an area, how they are transmitted, and how to prevent them.

Many diseases are caused by parasites like ticks and mites that enter the body, multiply, and spread disease organisms. If you know what carriers are responsible for a particular disease, you can then avoid sickness by avoiding the carrier.

NORMAL VITAL SIGNS	SYMPTOM	SIGNIFICANCE
Pulse Adult—60 to 80 beats/min. Child—80 to 100 beats/min.	No pulse Rapid/strong Rapid/weak	Cardiac arrest Fright, Hypertension Shock
Respiration Adult—15 breaths/minute Child—20 to 25 breaths/minute	Absent, labored, gasping Red, frothy blood	Airway obstruction, heart problem Lung injury
Eye Pupils Equal in size Constrict when exposed to light	Dilated Constricted Unequal Fixed, unmoving blank stare	Cardiac arrest, Unconsciousness, Fright Indicates drug use Head injury, Stroke Hysterical paralysis (scared to death)
Temperature 98.6 degrees F.	Skin—cool/dry cool/clammy hot/dry	Exposure to cold Shock Heat stroke, high fever
Skin Color Good indicator of body upsets for lightly pigmented people	Light pigmentation skin color—red —white —blue Dark pigmentation blue color around fingernails or inside lips of mouth	High blood pressure, heart attack, carbon monoxide poisoning, heat stroke Shock, heart attack Asphyxia, poisoning Oxygen problems
Mental Alertness	Disoriented, unresponsive, confused unconscious	Injury, body upset, lack of rest
Can Move and React Can voluntarily move body parts Can touch and feel	Unable to move legs Unable to move arms One side of body unable to move Numb feelings in arms, legs, body	Spinal cord injury below neck Spinal cord injury in neck Head injury, brain damage, stroke Spinal cord injury OR IN GENERAL, INJURIES AND/OR BODY UPSETS

PRESSURE POINTS FOR CONTROLLING ARTERIAL BLEEDING
USE PRESSURE POINT NEAREST THE WOUND, BETWEEN HEART AND WOUND

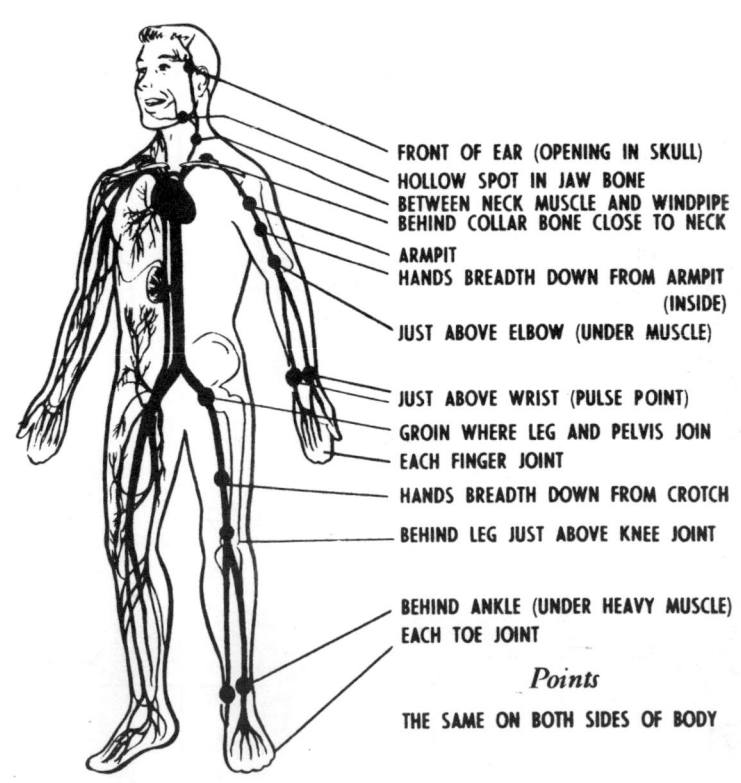

FRONT OF EAR (OPENING IN SKULL)
HOLLOW SPOT IN JAW BONE
BETWEEN NECK MUSCLE AND WINDPIPE
BEHIND COLLAR BONE CLOSE TO NECK
ARMPIT
HANDS BREADTH DOWN FROM ARMPIT (INSIDE)
JUST ABOVE ELBOW (UNDER MUSCLE)
JUST ABOVE WRIST (PULSE POINT)
GROIN WHERE LEG AND PELVIS JOIN
EACH FINGER JOINT
HANDS BREADTH DOWN FROM CROTCH
BEHIND LEG JUST ABOVE KNEE JOINT
BEHIND ANKLE (UNDER HEAVY MUSCLE)
EACH TOE JOINT

Points

THE SAME ON BOTH SIDES OF BODY

TOURNIQUET IS TO BE USED ONLY IN EXTREME BLEEDING OR AS A LAST RESORT

TYPE OF INJURY	SIGNS AND SYMPTOMS	TREATMENT
Bleeding—Severe Speedy and proper treatment essential Concentrate on keeping blood in the body	Arterial bleeding—pulses or spurs of bright red blood—WILL BE FATAL WITHIN MINUTES	Direct pressure to bleeding area—USE HAND. Apply compresses—use any material available—keep applying more compresses until bleeding stops. Use pressure points. Elevate wound, if no fracture. Apply cold packs, if available. Use constricting band (TOURNIQUET) only as a last resort and then leave on, do not loosen. (Loss of limb can be expected.)
Bleeding—Minor In general: clean, close cover	Abrasion/avulsion (removal of tissue)	Wash thoroughly—cover with sterile dressing—bandage.
	Puncture wound	Leave objects imbedded rather than cause further injury and bleeding by removal. Clean as thoroughly as possible. Cover with sterile dressing. Wait for evacuation.
	Lacerations/incisions	Clean wound with sterile compress and soap. Rinse with water, blot dry. If wound is open, close with butterfly bandages. Do not apply antiseptic solutions (iodine, merthiolate, mercurochrome) directly to wound.
Large Gash—Exposing Internal Organs	Self-evident	If organs are protruding, do not push them back. Keep moist with a saline solution (1 teaspoon of salt in quart of sterile water). Cover with sterile dressings—bandage applying minimum pressure, unless bleeding.

TYPE OF INJURY	SIGNS AND SYMPTOMS	TREATMENT
Bleeding—Internal	First sign is usually rise in pulse above 120. Dizziness when standing. Lungs—Coughing up bright red, frothy blood. Chest pains or outward evidence of injury. Stomach—Vomiting blood dark in color. Pain or signs of injury around stomach. Blood in urine or feces. Cold, clammy, pale skin. Pulse—rapid, weak. Breathing rapid. Excessive thirst.	Depending on type of injury, keep subject lying down or subject may be more comfortable partially sitting. Shelter, warmth. Apply cold packs to area. Give no liquids if evacuation is soon. Do not move extensively until evacuation.
Shock Failure of cardiovascular system to provide sufficient blood circulation. Normal reaction of body to injuries or emotional disturbances. May not occur until several hours after injury or emergency event.	Pulse—weak and rapid. Breathing—shallow, rapid, irregular. Skin—pale, feels cold and moist. "Cold sweat" is common. Eyes—appear vacant, may be dilated. Subject feels weak, faint, or dizzy. May be restless, frightened, or anxious. Often thirsty and nauseated. DEEP SHOCK—subject becomes quiet, slips into unconsciousness.	DO NOT WAIT FOR SYMPTOMS. EXPECT IT AND TREAT FOR IT. Lie subject down—shelter—insulate from ground—keep warm, not hot. Raise feet slightly, unless there is head injury. Treat injuries, relieve pain. If conscious, slowly administer warm salt solution (½ teaspoon of salt in quart of water).
Fractures—General Simple—no break in skin. Compound—bone has penetrated the skin.	Pain, swelling, deformity. Loss of movement and discoloration. Subject may have heard or felt a snap. Bone may be protruding through skin.	Avoid necessary movement, prevent further injury. Control bleeding. Immobilize the fractures—splint them "where they are" only if the limb is in normal anatomical position. Splint joints above and below the fracture. Reduce gross deformities—immobilize in a natural position. Reposition immediately after injury, if possible. Treat for shock. Keep compound fractures clean—wash with saline solution. Cover with sterile dressing.

TYPE OF INJURY	SIGNS AND SYMPTOMS	TREATMENT
Heart Attack	Chest pains, difficulty in breathing. Known condition. Swelling of feet and ankles. Blue color in lips, fingertips, earlobes, face. Weakness, dizziness, faintness. Skin may be cold and clammy. Chest pain may radiate to shoulders (especially left), neck, or jaw. Pain described as "crushing" or "pressure".	If conscious, get subject comfortable—loosen clothing—do not move. Keep warm. No liquids. Keep calm and rest. USE CPR IF BREATHING OR PULSE STOPS.

CARDIO-PULMONARY RESUSCITATION

1. IF VICTIM IS UNCONSCIOUS CLEAR AIRWAY

2. IF VICTIM IS NOT BREATHING
- Pinch nostrils closed.
- Inflate lungs rapidly 4 times.
- Open your mouth very wide. Make airtight seal over victim's mouth to prevent air leakage.
- Blow into victim's mouth to inflate lungs. Remove mouth and allow the air to escape.
- Check for pulse.

3. IF PULSE IS ABSENT
- Restore circulation by administering closed chest heart compression.
- Place heel of one hand 3 fingers from lower end of breast bone.

- Place victim on back on firm surface. Lift chin with fingers of one hand. With other hand, press forehead to tilt the head back.

- Look, listen. Feel to determine whether victim is breathing.

- Place other hand on top of the first hand, interlock fingers to assist in keeping fingers off of chest wall.
- Depress sternum 1½ to 2 inches. Repeat 80 times per minute.
- ONE OPERATOR. Inflate lungs 2 times after each 15 compressions.
- TWO OPERATORS. Inflate lungs once during the upstroke of each 5th compression, with no pause on compressions.

TYPE OF INJURY	SIGNS AND SYMPTOMS	TREATMENT
Head Injuries Concussion—stunning or jarring of the brain. Compression—swelling of the brain or hemorrhage within the skull. Check for spinal fracture, especially in neck.	Unconsciousness—the longer, the more serious. Pale face, cold skin, weak and rapid pulse, shallow breathing. Partial paralysis. Irrational, talking funny, or making nonsensical statements. Bleeding or fluid secretions from ears, nose, eyes. Unequal pupil size, unequal pupil response to light.	Control bleeding. Suspect neck injury. (Immobilize neck injuries on backboard.) Treat for shock. Place subject on abdomen, with head to one side, so blood/vomit can drain. Loosen clothing. Clear air passages. Face pale? Raise feet—lower head. Face flushed? Lower feet—raise head, especially if bleeding or secreting. Give nothing by mouth. Do not give stimulants.
Burns Generally, relieve pain by excluding air. Treat for shock.	1st degree burn—reddening of skin. 2nd degree burn—deeper burn with blistering. 3rd degree burn—tissue damage, charring, and cell damage.	Apply a salt free lotion. Do not apply lotions—immerse in cold water or cover with cold cloths. Cover with sterile dressing. Do not break blisters. Do not apply lotion or immerse in water. Cover with sterile dressing—wrap in plastic to exclude air, then ice packs to relieve pain. Treat for shock. Administer salty fluids (as much as tolerated).
Infection Blood poisoning is one of the final results of unchecked infection. Infection can extend from wound or to distant body parts through the blood (circulatory) system.	Wounds that bleed freely are less likely to be infected than wounds that do not bleed at all. Swelling, redness, pain, fever, sensation of heat, tenderness, pus, red streaks.	Treatment: Hot salt water or boiled water soaks and compresses. Penicillin or other antibiotics. Prevention: Avoid contamination of wounds and burns by unnecessary exposure. Do not contaminate sterile dressings. Use antiseptics on cuts and scratches. Use dressings that more than adequately cover the wound.

TYPE OF INJURY	SIGNS AND SYMPTOMS	TREATMENT
Eye Injuries	Pain, burning, redness, tearing, headache.	Do not attempt to remove foreign bodies from the eyeball. Turn head to one side, try flushing out object with water. Apply a light fluffy dressing. Foreign bodies may be removed from under eyelids. For chemical burns, flood eye with large quantities of water. (Don't wait for sterile water.) Prevent the injury from becoming worse. If only one eye is injured, bandage both to keep eyes stationary.
Dislocations Disturbance of normal joint alignment	Visual deformity, severe pain.	Check for pulse below injury. If no pulse, you must reduce dislocation or attempt return to normality, by pulling slowly, steadily, forcefully, and guiding the misplaced bone back into position. (Loss of limb will occur if no circulation.) Do not attempt to relocate injury if aid or rescue is near.
Sprains Stretching of damaging of tendons, ligaments, tissues, and blood vessels surrounding a joint.	Pain, swelling, discoloration, tenderness.	In doubt? Treat as a fracture. If sure—elevate injured area—use slings—apply ice or cold compresses to reduce swelling and pain. Wrap joint if you must walk. Do not remove shoes.

TYPE OF INJURY	SIGNS AND SYMPTOMS	TREATMENT
Back Injury Fracture of spine or neck	Pain, shock, paralysis. Acute pain at point of fracture. Broken back—paralysis of legs, feet, toes. Broken neck—paralysis of fingers.	Minimize shock—prevent further injury. Do not move unless absolutely necessary. Do not bend, twist, or move head and body. IMMOBILIZE—keep subject lying down.
Snakebite	Within 30 minutes, pain, swelling, discoloration, weakness, dizziness, faint pulse.	Apply suction to bite without incision. Use mouth, if no suction cups. Spit out venom, rinse mouth often. If pain, swelling, numbness, discoloration occurs, then, place constricting band above bite. Constricting band should be snug but not restrict pulse below band. Apply others if swelling extends. Wash skin around wound. Sterilize knife or razor blade. Make linear incision through each fang mark, 1/8" to 1/4" long and no more than 1/8" deep. Be careful not to damage veins, nerves, or muscles. Apply suction to incisions. Wash wound, bandage. Treat for shock. Watch breathing.
General Wounds		Expose wounds—remove foreign bodies from surface. Do not probe depths of wound. Control bleeding. Prevent infection, apply antiseptic, apply large dressing. Treat for shock. Irrigate wounds with sterile water, do not scrub.
Ears	Earaches may indicate infection or perforated eardrums.	Do not put anything into ear. Do not probe for foreign objects. Apply a light dressing.

Insects

Insects can be a source of danger and discomfort far out of proportion to size. Their greatest threat is transmitting weakening, often fatal, diseases through bites.

Mosquitoes

Mosquito bites can lead to death in several ways. These pesky insects are found everywhere, but are most dangerous in the tropics where they carry malaria, yellow fever, dengue fever, encephalitis, and filariasis. In northern climates, their mass numbers can drive an animal or a human beyond limits of endurance.

When traveling in mosquito country
- stay on high ground away from swamps.
- use a mosquito net, headnet, gloves, or anything available. Smear mud on exposed skin if nothing else is available.
- keep clothes on, especially at night.
- tuck pants into socks or shoes.
- use mosquito repellent. On clothes it will last weeks. On skin it wears off in hours.
- take anti-malaria tablets if necessary and available.

Flies

Like mosquitoes, flies vary in size, breeding habits, and possible danger or annoyance. Most measures effective against mosquitoes also work for flies.

Fleas

These small wingless insects can transmit plague to man after feeding on infected rodents. If necessary to eat rodents in plague areas, hang them up as soon as they are killed because fleas leave a cold body. Use flea repellent powder on clothing and wear tight-fitting boots.

Ticks

Ticks are found throughout the world. They are carriers of tick relapsing fever and tick typhus. Two major types are hard (or wood) tick and soft tick.

Check body thoroughly at least once a day in tick country. If one is found on the body, apply moderate heat with a burnt match or hot pin to make the insect release and back out of the skin, then disinfect thoroughly.

Mites, Chiggers, and Lice

These very small insects are common all over the world. Chiggers are immature mites which bore into skin, causing itching and discomfort. These bites can make people ill. Some chiggers can transmit scrub typhus, and others cause various skin diseases with possible secondary infection from scratching.

Louse bites should not be scratched because one can spread louse feces into the bite, possibly causing epidemic typhus and louse-borne relapsing fever. If louse powder is not available, boil clothing to remove lice, or expose clothes to direct sunlight for a few hours. Wash, preferably with soap, sediment, or sand from stream bottoms and check hairy body parts for lice.

Spiders

Most spiders are not particularly dangerous. Even tarantula bites are rarely serious. Black widow bites, and those of tropical members of the same family, cause severe pain, swelling, and possible death. These spiders are marked with white, yellow, or red spots. A bite by one of these can cause abdominal cramps which continue intermittently for a day or two. The pain is similar to acute indigestion.

Scorpions

Scorpion stings are painful but seldom fatal. Some larger species are more dangerous and can kill. Scorpions are widely spread and can be dangerous as they hide in clothing, shoes, and bedding. Shake out clothes before dressing. Cold compresses or mud are helpful on scorpion stings.

Centipedes and Caterpillars

Centipedes are numerous in semi-tropical areas and some larger ones give painful bites. They seldom bite humans unless trapped when they have taken shelter in clothing. Centipedes and caterpillars can cause inflammation and itching when touched.

Bees, Hornets, and Wasps

A swarm of bees, hornets, or wasps can be dangerous — even fatal. Avoid nests. If attacked, run through dense brush or undergrowth and let twigs beat off the insects. Going underwater also discourages these insects. Cold compresses or mud are helpful on bee, hornet, and wasp stings. Some people are allergic to bee stings, and must be rushed to a doctor for counteracting injections.

Shelter

Without shelter, humans can only survive within a narrow body core temperature range. The ability to maintan the body's 98.6° F. temperature, through a knowledge of body shelter management, is one of the keys to maintaining life. Shelter consists of anything that protects the body from temperature, weather, or any other life-threatening force. It may or may not have a heat source, though in colder climates, some form of heat is desired for additional comfort. In an emergency, find protection from storm, wind, etc. as soon as possible.

> **Parts of the aircraft will be your fastest, most available resources for shelter and insulation.**

Do what animals do and burrow in or crawl under foliage. Burrow out a small shelter very rapidly in forest areas. Fire and light are pleasant, but starting a fire in rain or high wind may be futile and waste body heat and wet clothing, which could be retained in a burrow.

> **Seek immediate shelter and stay dry.**

Whenever a situation requires immediate body protection, a shelter can be improvised from what you have with you or around you. The type of workmanship and materials will determine a shelter's efficiency. Efficient shelters need little or no external head to maintain body warmth. Adequate body shelter can be improvised from natural and/or manmade materials.

> **Shelter is only limited by**
> - **your imagination**
> - **immediate physical needs**
> - **materials available**
> - **body energy**

Shelter is a supplement to clothing, and a properly built shelter should encapsulate the body from wind, wet, and cold. Through the lack of adequate body shelter or clothing, limited supplies of body heat, energy, and coolant are quickly lost. Remember, a sleeping bag is a shelter composed of two layers of relatively weatherproof fabric separated by insulating material. Heat produced by the body is retained inside the bag even when outside temperatures drop below freezing.

Efficiency of sleeping bags are improved by added ground insulation, added filler (insulation) in the bag itself, and overhead and side protection (tent or tarp) from wind and rain. Keep this sleeping bag principle in mind when building emergency shelters.

> **Shelters should always be kept small, dry, and well insulated. Less body heat will be needed to warm it and warmth will be retained longer.**

Shelter can be best defined as "anything that retains or dissipates body heat or protects the body from dangerous elements."

Build it
- simple and small, no bigger than what is absolutely needed for body protection.
- within the limits of your physical ability.
- within the limits of your body's energy and coolant.
- to minimize body heat loss or gain through conduction, convection, evaporation, and radiation.

Checklist: Considerations for Shelter Building

Shelter Location
The aircraft provides the best immediate shelter. Stay with or near the aircraft. Check the area for possible or future dangers. Areas to avoid are dry stream beds, avalanche chutes, rock fall areas and dead trees. Also, pay attention to insects or animals present.

Usable Materials
Determine the type of shelter needed. Remember that function is more important than shape. Three factors basic to all shelters are a roof, walls, and floor.

> - *Many aircraft parts can be quickly utilized.*
> - *Be careful in using metal parts in extreme heat or cold.*
> - *Metal will rapidly increase or decrease body heat through conduction.*

Find something nearby already partially built; a hollow tree trunk or a fallen tree can, with little effort and aircraft insulation materials, be turned into a comfortable shelter. In areas with no trees, build a shelter with aircraft nonconductive materials, dirt, brush, etc.

An important principle to remember in using natural materials is "as it grows in nature, so place it on your shelter." If used as it grows, it will be stronger, and more effective.

Basic principles of heat transfer (body heat loss or gain) and thermal conductivity (insulation) must be understood.
- Thermal condutivity and heat transfer is the measure of any substance's ability (or inability) to conduct heat. Comparisons of example substances are shown in the following table.
- A lower value indicates a better insulator.
- In terms of survival, remember that the human body will lose or gain heat anytime it is in contact with anything that is cooler or warmer than body temperature.

TABLE OF THERMAL CONDUCTIVITY
Reducing this table to air as the base produces these relative values

Air	5.7×10^{-6}	1.0	Air
Wood	2×10^{-5}	3.5	Wood
Glass	2×10^{-4}	7.0	Glass
Ice	4×10^{-4}	7.0	Ice
Lead	8.3×10^{-3}	145	Lead
Steel	1.1×10^{-2}	1930	Steel
Aluminum	4.9×10^{-2}	8600	Aluminum
Silver	9.9×10^{-2}	17400	Silver

Insulation Sources

It is essential to insulate the floor to prevent heat loss or heat gain, using *nonconductive materials from the aircraft,* or natural materials: bark, deadwood, thick boughs, grass, or leaves.

Insulation is anything that will provide dead air spaces between you and the ground and environment. Keep all insulation materials dry and fluffy. Compressed materials lose trapped air spaces. Beware of any materials that conduct heat rapidly.

Fuel Sources

Is there adequate natural fuels for a fire? Should the shelter be constructed to allow a fire inside or near the opening? Is fire really needed? Remember, a small, cramped, slightly uncomfortable shelter may be better as it will minimize heat loss.

Wind Direction

Check wind direction and exposure. Wind will generally blow down canyons at night and up canyons during the day. The shelter entrance should be 90 degrees to the prevailing wind.

Water Sources

Is the location in close proximity to water Avoid locations near creeks, rivers, lakes, etc. as it will be damper and cooler.

Signalling

Is there an open area for signals? What kinds of materials could be used? How visible is the aircraft? Can you make it more visible? What materials from the aircraft could be used for signalling?

General

Determine what is needed . . . roof, walls, and floor to protect the body from the environment. Adequate shelter can always be constructed from materials at hand. *Think and improvise.*

Improve what is already available. Enlarge the opening so it is possible to get in and out easily. Try to make enough room to stretch legs.

Can light be seen through holes in the walls or roof? Dirt and grass can be used for chinking cracks or making the wall base.

Once the shelter is improved as much as possible with materials at hand, decide what else is needed. Find materials to build it.

Review this list of shelter considerations. Has anything been overlooked?

Hot Weather Shelter Considerations
- Shade from the sun is the most important daytime shelter requirement. Any materials or objects that produce shade will be an asset.
- Moving air (convection) will be helpful. Beware of still air canyons.
- Seek shade, underground or above ground. Air temperature 18 to 24 inches above or below the ground surface is usually 30 to 40 degrees cooler.
- Keep clothes on and head covered.
- Do not stay in a closed metal structure (aircraft) without adequate ventilation. Temperatures can become intolerable and can cause body heat problems.
- At night, when desert air cools considerably, better shelter and fire might be needed. Work during cooler periods to improve shelter and general situation.
- Days should be spent resting in shaded, elevated shelter, with clothes on.

Cold Weather Shelter Considerations
- Forested areas can provide overhead shelter, fuel for fires, and materials for insulation. A "downed tree shelter" is the easiest natural shelter of all. It already has a roof, floor, and one side. If there is snow, a "tree pit shelter" may be readily available and it is over half made.
- On wind blown prairies or exposed ridges, scrape out a trench in the snow or ground. Pile rocks or other materials for a wind break. Line the trench with limbs, grass, bark, aircraft materials, extra clothes, or anything that will provide insulation. Put snow or dirt on top of a framework . . . crawl in . . . and close up.
- Snow in various forms is a good insulator and gives adequate wind protection. It is hard to work with, requires a degree of practice and skill, and takes considerable energy. Digging a snow cave takes a long time and requires a tool. Digging into a snow bank may use less energy. Work at a pace to minimize sweating. Keep clothes dry. Snow shelters must have adequate

ventilation. A candle provides light, some heat, and is also an excellent indicator that oxygen is running out (provide better ventilation). Avoid overheating the inside of a snow shelter. If it melts, you will get wet. If ice builds up, the shelter will become a conductor of heat rather than an insulator.
- If temperatures are extremely cold (10 degrees F. and colder) long term shelter should not be taken in the aircraft. Metal parts become heat conductors and very rapidly deplete body heat.

Checklists for Each Phase of the Emergency Environment

Preparing Today for Any Emergency (Pre-emergency Phase)

- [ ] Practice emergency landing techniques, 180 degree turns, power off at altitude. Establish your emergency checklists.
- [ ] Chop the power a number of times in your aircraft and practice setting up proper trim and glide.
- [ ] Check the batteries in your ELT as well as the mounting and antenna cable. Move it, or have it moved if unsatisfactory.
- [ ] Take a first aid course. Keep current on the latest first aid procedures. Learn about survival first aid as opposed to standard first aid.
- [ ] Take a good comprehensive emergency preparedness (survival) course which covers a full spectrum of skills and orientation.
- [ ] Develop skills for fire making, shelter construction, emergency first aid, water procurement, and signalling.
- [ ] Plan and build a comprehensive emergency kit that will serve your emergency needs.
- [ ] Increase your comfort zone by experiencing new things.
- [ ] Be conscious of what your body is telling you. Learn how to properly manage your body in emergency environments.
- [ ] Always file a flight plan and, whenever possible, back it up with verbal intentions.

Initial Response (Emergency Phase)

- [ ] Maintain control of the aircraft and immediately go through instrument check and restart procedures. Should be methodical and in accordance with **your** pre-established emergency procedures.
- [ ] Radio MAYDAY and/or transmission to the nearest ground facility. Successful receipt of a distress signal is often times dependent on altitude.
- [ ] After radio transmission, activate ELT from cockpit if possible.

- ☐ Set up the aircraft in the maximum glide ratio attitude.
- ☐ Assess the terrain for suitable landing areas and pick one.
- ☐ Instruct passengers on exactly what to do, ie., the use of travel pillows, proper seat positioning, securing items inside cockpit, opening door slightly over water, etc. (Be positive and reinforcing in your instructions.)
- ☐ Check seat belts and shoulder harnesses to see that they are secured tightly.
- ☐ Look for a more specific landing sight as altitude decreases (clearings, flat terrain, smaller trees, uphill slope configuration in steep terrain, roads, etc.)
- ☐ Fly the aircraft until it completely stops and never try to stretch your glide.
- ☐ Use non-essential parts of the aircraft to slow down, absorb energy and maintain the integrity of the cockpit.
- ☐ Monitor air speed and avoid high sink rates as a result of stall.
- ☐ Switch power off to all electrical systems at the last moment.
- ☐ After the aircraft has come to a complete stop, try to get away from it until it cools. In many cases this may not be possible.
- ☐ Administor self aid or first aid to passengers as soon as possible when necessary.
- ☐ Try to maintain a positive mental attitude with specific activities that are objective oriented.

Life Support (Post Emergency Phase)
- ☐ Maintain a positive mental attitude with constructive activities.
- ☐ Provide self aid or first aid for others with the idea that second aid or more definitive care may be called for in the situation.
- ☐ Assess the situation and observe the surroundings. Make a mental inventory of what you have to work with, both in the aircraft and in the environment.
- ☐ Obtain your survival kit and other possessions and assess your shelter needs. Can the aircraft be used for shelter or be modified for shelter, adequate insulation, etc.
- ☐ Plan for a period of shock, rest and recovery.
- ☐ Check radio and ELT and the battery for each.
- ☐ Do not over-extend yourself physically. Conserve the energy that you have available already in your system.
- ☐ Gather all of the materials that you have for signalling, ie., pyrotechniques, bright colored tarps, mirrors, etc.
- ☐ Assess aircraft visibility from the air and determine whether it can be improved.
- ☐ Determine the feasability of fire with wood fuels and aircraft parts. Will the effort be worth the energy output, ie., is it raining, snowing, or blowing, etc.?

- [ ] Maintain body temperature with hot or cold liquids and insulation in accordance with hot or cold environments.
- [ ] Make plans as to exactly what you are going to do and follow through to accomplish those objectives. Re-establish goals and objectives which are commensurate with the situation.
- [ ] Assess needs, water, etc. and potential sources for improvising those needs.
- [ ] Re-assess priorities and needs as the situation develops.
- [ ] Conserve energy and begin to establish as many ways as possible to signal your distress.

Rescue Phase (Post Emergency Phase)

- [ ] The rescue phase should have begun before the flight began by the filing of a complete flight plan and letting others know your exact plans and contingencies.
- [ ] MAYDAY and call to nearest ground facility must be done at altitude. Give situation, approximate location, and other pertinent information.
- [ ] Activate ELT and double check to assure correct operation. Improvise antenna if necessary when on the ground.
- [ ] Do not forget to try the radio if the battery is still intact.
- [ ] Establish ground signal or fire and stay with the aircraft.
- [ ] Continue to assess the aircraft's visibility from the air and continue to improve, ie., remove snow or debris from top of the aircraft.
- [ ] Establish a convenient place to have all pyrotechniques and other hand held signal devices ready for use.
- [ ] Remember the following signalling tips:
  - There are very few straight lines in nature.
  - Use well defined right angles and contrast with background.
  - Color has a great deal to do with visibility.
  - Correct size ratio for ground signals is 6:1.
  - International distress signal is three anything.
  - Search personnel will be looking for anything out of place and any movement.
- [ ] Evaluate your efforts and determine their effect.
- [ ] Maintain a positive mental attitude. Set realistic goals and use self talk.

Clothing

Flying poses unique problems in regards to the wearing of proper, adequate clothing. Far too many lightly dressed pilots and passengers become accustom to flying over various hostile environments, sitting in the cockpit's regulated "green-house" environment. They give no thought to the additional clothing requirements necessary for survival should an emergency landing occur. Search and rescue statistics show that the average pilot and passengers were not wearing or carrying the clothing necessary to survive in the environment that they suddenly found themselves in.

Before staring out, ask yourself "Is what I plan to wear and carry adequate for the expected weather and environments that I will be flying over?"

> **Humans maintain a constant body temperature by regulating the type and amount of clothing worn.**

Unfortunately, the clothing produced by modern society often does not provide adequate body protection. Fads and "in" fashions dictate what clothing to wear. Even most outdoor and work clothing is designed for fashion rather than for retaining or dissipating body heat.

An important aspect of emergency preparedness is the clothing worn on the body. Know the purpose of clothing, the various types of clothing materials, and the characteristics of clothing materials when they become wet. The ability to function and maintain comfort in natural environments is directly related to these factors.

Personal Protection

The human animal in natural state is suited only for tropical, semi-tropical, or desert areas of the globe. The nude human body cannot withstand temperature extremes of any great variance. Through history, as man moved away from warmer climates, he found body covering or shelter necessary if colder areas were to be explored and settled. From primative beginnings with animal skins and plants, man progressed to sophisticated development and production of clothing to cope with the rigors of nature.

Clothing is shelter close to the body, intended to help maintain delicate internal body temperature balance. Clothing traps radiated body heat into dead air space and helps eliminate convection currents which would remove heat. Clothing's effectiveness is judged by the rate at which the body's radiated heat passes across dead air spaces between skin and clothing barriers.

Heat transfer depends on several factors. First is thickness of dead air spaces. If skin touches clothing, conduction transfer is rapid. For cold, still air conditions, thick layers of dead air spaces found in down, or other fluff-filled clothing is needed. The second factor is amount of air which passes through material weave causing heat loss through convection and evaporation. Large knit, wool sweaters are fine in still air, but lose heat rapidly when wind blows. Third is moisture factor. Heat loss through wet clothing is very rapid since thermal conductivity of water is 240 times as great as still air. Wet clothing extracts body heat nearly 200 times faster than dry clothing.

Heat transfer through clothing takes place by
- conduction through metallic objects such as buttons, zippers, nails in shoes, and buckles.
- conduction and evaporation through wet material.
- convection, radiation, or evaporation through space between fibers of the weave.

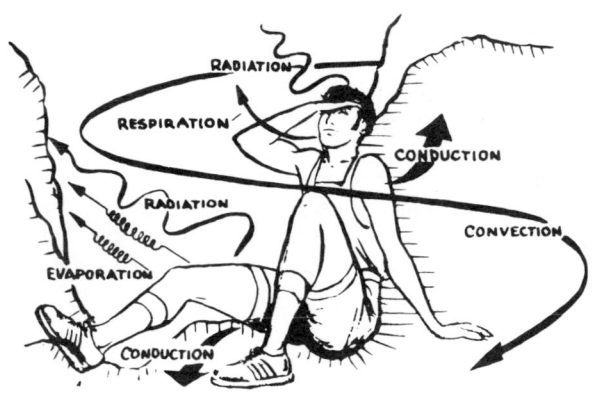

Mechanisms of Body Heat Loss

Conduction
- the primary cause of body heat loss.
- a transfer of heat occurs anytime the body touches something colder than body temperature.

Radiation
- the body continually radiates heat from all exposed areas of skin.
- 50% of body heat is lost through the head area.

Convection
- air currents blow the radiated heat away faster than it can be produced.

Evaporation
- body moisture can wet clothing, compounding conduction.
- heat is lost through the evaporation process.

Respiration
- in cold environments, cold air enters the body and leaves as warm air. The body loses heat warming the cold air.

Materials
Cotton
In warm or moderate climates, cotton provides excellent properties for protection. It affords insulation against cold when used in the layer system while it remains dry. Cotton's principal danger lies in water retention when wet from rain, snow, or perspiration. It dries slowly because of complete saturation of fibers. Its heat conduction properties when wet nearly equal complete immersion in water. It also wicks water rapidly from damp or wet areas to dry areas. Poplin is a generic term referring to tightly woven fabric made of fine cotton yarn. Ventile cloth is poplin originally manufactured to protect downed British fliers in the North Sea. It breathes and temporarily resists water absorption. As the material absorbs water, it swells and closes off spaces between the yarn fibers.

Wool
The principal advantage of wool over other fabrics is its ability to insulate when wet. The fibers are naturally curly and trap air in tiny pockets between fibers. Wool exhibits little or no wicking action. Raw wool contains lanolin which repels water. The disadvantage of raw wool is a distinctive odor. Successive washings eventually remove lanolin while dry cleaning removes it much faster. To keep wool garments waterproof, they should not be washed often. Search and rescue statistics prove people dressed in wool are more apt to survive than those dressed in cotton. However, the loose weave of woolen fabric leaves much to be desired in a windstorm. A very close weave wind-resistant material should be used for outside clothing under these conditions.

Nylon
Although stronger than cotton for equivalent weight, abrasion resistant, quick drying, and long wearing, synthetic threads are completely impervious to penetration both by water and water-repelling compounds. This presents difficulty in two respects. First — perspiration stays on the surface causing dampness. Second — waterproofing compounds do not unite with fiber as with cotton, and therefore must be "painted" into material to fill in between threads. Untreated nylon is excellent for wind-only protection. Ripstop nylon is generally a lighter grade nylon with intermittent heavier threads which stop a rip from running throughout the material. It is often used for wind-resistant clothing.

Blends
To produce better, stronger, and more durable materials, the clothing industry mixes various fabrics to form a blend. Small

percentages of nylon are often mixed with wool yarns to produce socks containing properties of wool with the increased strength and durability of nylon. Sixty-forty fabric is a weave with nylon threads going one direction and cotton going the other. The nylon provides strength and the cotton swells at first contact with water. Although not waterproof, it is somewhat water repellent. Sixty-five/thirty-five fabric is a blend of cotton and Dacron polyester. It has greater strength than sixty-forty, and accepts water repellency treatments more evenly. Other blends are available and most take on characteristics of the greater portion of fabric present. Synthetics in general have poor insulation qualities under wet conditions. There still is no substitute for wool in natural environment.

The Layer System

The only effective way to regulate body temperature is through the layer system. Each human is a heat-producing organism depending on clothing for protection from temperature extremes. Every individual metabolism is different as is every tolerance level and comfort zone. Because of this, every individual should dress according to planned activities and exertion. How much clothing is enough? What is enough for one may not be enough for another. People who constantly live and work outdoors have developed the layer system of dressing—wearing a number of easy-on-easy-off layers of clothing rather than one large garment such as a parka. Slowly shed a number of layers of clothing as physical exertion continues and more body heat is produced. Clothing layers can easily be re-donned as heat production tapers off. This method of dressing is dependent on several thin layers of insulation and trapped dead air instead of a single massive layer. Ventilation with rolled up sleeves and unbuttoned front extends use of each layer.

Improvised Clothing

It may be necessary to improvise clothing from resources immediately available. In cold it might be imperative to get more insulation. Stuffing from aircraft seats, newspapers, grass, leaves, or pine needles, provide additional dead air space to supplement clothing. Fashion sewing needles from wire, paper clips, toothpicks, key rings, and even eyeglasses. Material to make boots, shoes, or additional outside clothing can be sewed from seat covers, floor mats, maps, rugs or other available items. Think! Your brain is your best tool. You are only handicapped by imagination. Many emergency situations away from controlled heat shelter require improvised clothing.

Survival Tips on Clothing
Cold
- When your feet are cold, put on a hat.
- Do not over-heat so that clothing becomes wet by perspiration.
- Use the layer system.
- When sleeping in harsh cold conditions, arrange dry spare clothing around neck and shoulders with padding and insulation added to each kidney region. These areas are more susceptible to cold.
- If a person falls in very cold water, roll in snow to blot out moisture immediately. Wiggle toes and get into dry clothing as soon as possible.
- Dry wet clothing by allowing it to freeze and then beating ice crystals from fabric.
- Wear darker clothing in winter to absorb sun heat energy.
- Clean clothing allows proper ventilation through clothing layers. Dirty clothing inhibits ventilation and causes moisture buildup in clothing layers.

Heat
- Light colored clothing reflects heat and helps maintain temperature balance.
- Body water level must be maintained. Since convection cools the body by evaporation or perspiration, reduce evaporation rate to cut down water loss. Leave clothes on to reduce convection currents reaching the skin. This may seem uncomfortable because of the increase in skin wetness and apparent increase of temperature, but the body stays cooler and water loss is substantially lower. Long flowing robes worn in deserts demonstrate this concept. Clothing should not be discarded in a hot environment.

Firecraft

Fire building is a skill that is only mastered by practicing in a variety of conditions with various materials. Proficiency requires a basic knowledge of why a fire burns and which fuels are best. During emergencies, fire can be very important as it can provide warmth, dry clothing, give light, cook foods, provide a signal, boost morale, and purify water.

> *Your emerency preparedness kit should contain several methods of fire starting.*

Ask yourself: "Do I really need a fire? What do I want it to provide? Is it worth the cost of energy expended to build and maintain it?" The type and size of your shelter, quality of materials used,

and workmanship will determine your true need of fire and warmth. The smaller, drier, and better insulated your shelter is, the less heat is lost, and the less need you have for fire.

Remember that
- the better the shelter, the less need for fire.
- inadequate clothing will require adequate shelter.
- inadequate clothing and inadequate shelter will require fire.

> **Reasons for firecraft failure**
> - *impatience*
> - *inexperience*
> - *poor selection of tinder, kindling and fuel*

Location of a Fire

Look for a location with little or no burnable materials that is away from trees. Sandy or gravelly soil without grass or roots is best. The area should be protected from the wind, near potable water, if possible, and near fuel supplies.

Make a fire circle to insure that the fire does not spread out of control. The fire circle should be five to six feet in diamter on sandy or gravelly soil and scraped down to mineral earth that is free of duff, grass, and roots.

A circle of *dry* rocks can contain the fire and is helpful for cooking. *Caution: river rocks contain moisture and can explode when heated.*

Never build a fire
- near dry, flammable materials.
- in very dry grasslands.
- under overhanging branches.
- in a very resinous (pitch smelling), dry forest.
- under a snow laden tree.
- directly on the snow.
- on or near wet rocks.

Requirements for Fire

Only three factors are necessary to start a fire. These are heat, fuel and oxygen.

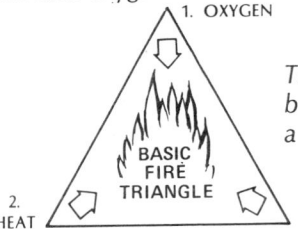

The most common mistake in fire building is improperly placing wood on a fire — getting a smudge pot effect.

Initial Heat Sources
Friction
- Matches, strike anywhere and waterproofed (use with candle or fuzz stick).
- Bow and drill, rubbing sticks together, hand drill, fire saw, plow board.
- Note: There are many ways to start fires using friction. There are many primitive living skill books that will show you how. Acquiring these skills takes alot of time, practice, and energy. These skills are great confidence builders and back-up systems. It is far better to carry matches and several types of spark fire starters.

Sparks
- Flint and Steel
- Aircraft battery
- Cigarette lighter
- Metal match
- Magnesium fire starter
- Scraping knife or metal against a hard rock
- Note: Check sporting goods stores and recreation stores for many types of spark producing fire starters.

Sunlight
- Magnifying glass
- Any convex lens from a camera or binoculars

Fire Materials — Tinder, Kindling, Fuel

All fire materials should be gathered before attempting to start a fire. Try to use the driest possible materials gained from hollow logs, dead stumps, dead limbs on living trees, or dead twigs and limbs found off the damp ground.

In very wet weather, dry wood can be found in the center of old stumps and the center of dead standing trees. Dry twigs can usually be found near the trunks of larger trees. Old burned out or broken off stumps (and any spires sticking up out of the stump) usually indicate high pitch concentrations ideal for buildling fires.

Gather Three Piles of Fire Materials

1. Tinder is anything that will ignite at a very low temperature, with a spark, small flame, or other heat source. *Tinder must be dry.*

   Consider the following:
   - dry grass, crushed
   - cotton or scraped cloth
   - fine, dry wood shavings
   - dry cattail fuzz
   - bird down

- pitch wood, fine shavings or powdered wood
- lint on clothing or in belly button
- steel wool
- petroleum products
- dry, reddish pine needles
- seed down
- shaved sticks
- hair
- paper
- inner bark of cedar or birch
- dried moss or lichen
- candle
- oil or gas soaked cloth

2. Kindling is the initial fuel stage. Find something that will ignite easily from the tinder. It needs to be small in diameter, broken, split, or shaved to increase flammability.

Consider the following:
- finely split wood
- fuzz stick, carved with jack-knife
- dry, dead branches and twigs

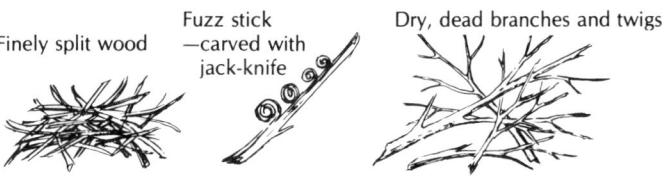

Finely split wood   Fuzz stick —carved with jack-knife   Dry, dead branches and twigs

3. Fuel is anything that will burn for an extended period of time. Generally it will not burn from the initial flame or spark and requires high temperatures for continued burning.

Consider the following:
- dead wood
- dry peat
- coal
- dried dung
- rubber
- green wood
- animal oil or fat
- bundles of grass
- pitch
- petroleum.

Caution: Many manmade products will emit poisonous vapors when burned. Some explode.

Starting the Fire

The tinder should be arranged so that heat from the tinder or flame rises through the maximum amount of kindling. A teepee shape works best. When the kindling begins to burn well, add larger pieces of fuel. Impatient beginners usually smother the kindling flame by adding too much fuel, which cuts off the draft and oxygen.

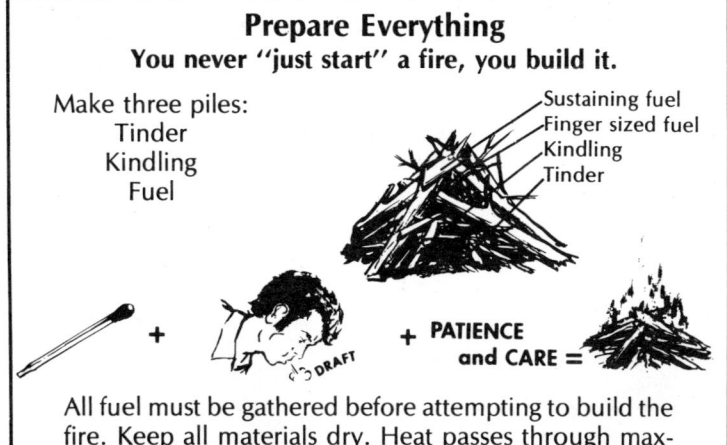

Prepare Everything
You never "just start" a fire, you build it.

Make three piles:
Tinder
Kindling
Fuel

- Sustaining fuel
- Finger sized fuel
- Kindling
- Tinder

+ DRAFT + PATIENCE and CARE =

All fuel must be gathered before attempting to build the fire. Keep all materials dry. Heat passes through maximum amount of fuel.

Fire Without Wood Fuel

A metal can with heavy oil and a cloth wick can serve as an emergency stove. Oil soaked rags or shredded upholstery in a shallow pan of oil will also burn. Fill a can one-third full of dry dirt or sand. Make draft and smoke holes on the sides. Saturate with gasoline or diesel. Wait a few minutes, ignite with a *tossed* match or spark. Stir occasionally.

Uses and Types of Fires

Platform Fire

On sheet metal or green logs. Use on ice, snow, or mud.

Log Cabin Fire

Provides a hot bed of coals; good for cooking and drying. (White Man's Fire.)

Star Fire

Conserves fuel, eliminates need to cut big chunks into short lengths.

Tepee Fire

Concentrates heat — For general cooking, heating and drying.

Long or Trench Fire

Excellent for larger groups, gives light, warmth, and at the same time cooking on coals. Also conserves fuel.

Firecraft Tips
- Conserve matches whenever possible by lighting a candle or fuzz stick.
- Green wood will burn if finely split.
- You can find dry wood for fire starting in the center of standing dead trees or wood. Small dry twigs are often available near the base of green trees.

- If you have no cutting tool sufficient for splitting wood, whittle a hardboard wedge, and drive it into weathering cracks in the end of the wood.
- A reflector (metal, rock, wood) on one side of a fire makes it more efficient. For a firesite with a large rock surface or at the base of a cliff, do not use the rock as a reflector. Build the fire far enough away from the rock so that you can sleep between the two. The rock will provide warmth on one side, while the fire warms the other.
- Fire supplements the body's heat producing mechanism. In colder weather, several small fires built around you heat better than a single, larger fire.
- Cooking fires should be walled-in to concentrate the heat.
- If a fire must be built in deep snow, built it on a platform of crisscrossed green logs to keep the fire from sinking.

Signalling

Emergency signalling is a skill needed to make rescuers aware of your distress and location. Properly used, signals will make you more visible. The sooner you effectively signal, the sooner you may receive assistance.

> **Signalling is you taking an active part in your rescue.**

Signalling is the common sense knowledge of what to use, when to use it, and how to communicate to others without the spoken word.

> **Contrast to your environment is the key to effective signals.**

Be ingenius and you will be noticed.
Consider the following:
- fires
- smoke
- lights
- shadows
- movement
- flags
- dyes
- straight lines
- right angles
- anything out of place

> *Your aircraft, with a working ELT may be your best signal. Stay with the aircraft. Make it more visible.*

Your emergency preparedness kit should contain provisions for signalling. Though the aircraft itself, an ELT, and a fire may be adequate, the more signals you make, the better your chances are of being seen.

Consider carrying the following:
- flares
- dyes
- colored signalling panels
- lights
- rockets
- mirrors
- whistles
- canned smoke

Standard Signals

The international distress sign is a series of three repetitions of any signal, i.e. three loud sounds, three fires set in a large equilateral triangle, three flares, etc. Voice distress is **Mayday, Mayday, Mayday** or **S.O.S.** Acknowledgment by rescuers is two repetitions of any signal.

Smoke Signals

To be effective smoke signals need to contrast with the environment. Such as dark smoke against snow, white smoke against dark turf, dark smoke against overcast skies, or light smoke against clear skies.

For black smoke add the following to the fire:
- engine oil
- rags soaked in oil
- pieces of any rubber product
- plastic
- pitch
- pitch wood

For white smoke add the following to the fire:
- green leaves
- moss
- ferns
- green boughs
- some water, wet cloths

In some environments, an evergreen torch can be an extremely effective signal. Select a small evergreen tree with dense foliage,

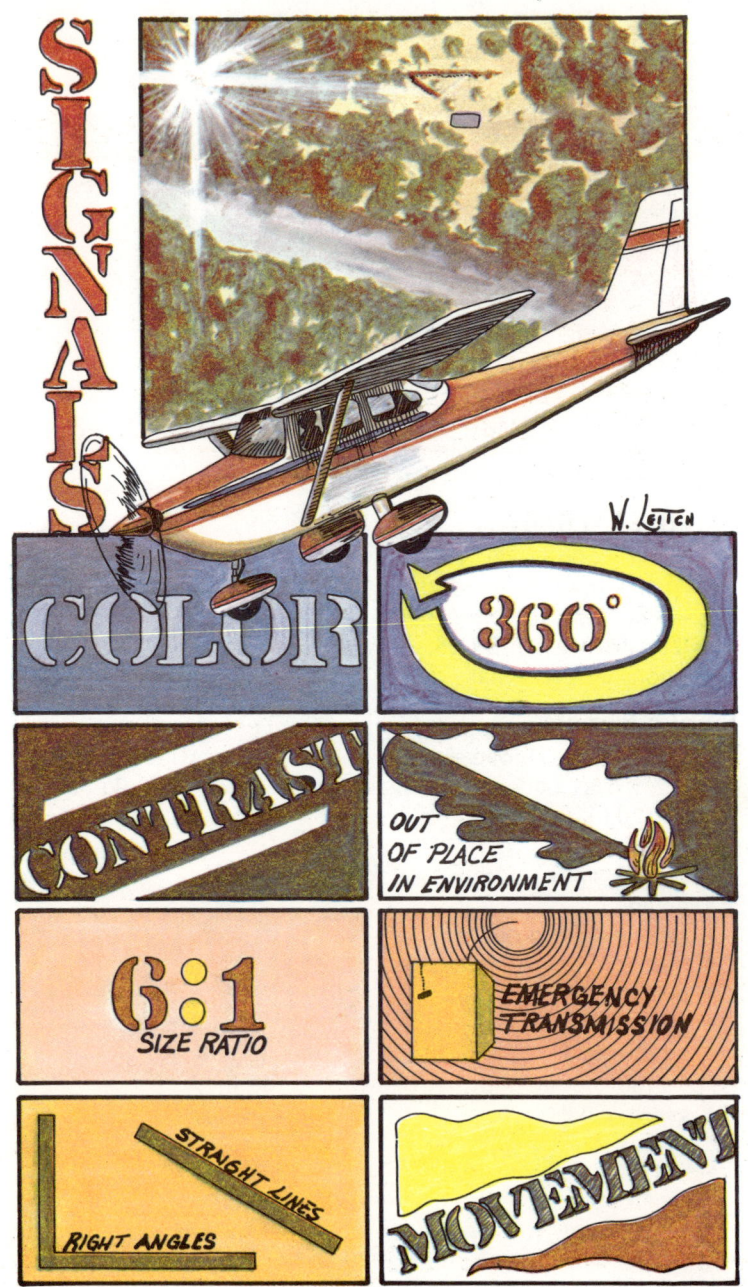

cut it down, and move it to the center of a large clearing or some other location where fire is not likely to get out of control. Place dry tinder in lower branches. If an aircraft is heard or seen, light the tree. In most cases this signal will punch smoke from 1200 to 1500 feet straight up.

Signal Mirror

Practice mirror signalling technique in advance — aircraft pass over very rapidly. If a person is not ready when an aircraft appears, it will be gone without seeing the signal.

Mirror flashes or reflections have saved many lives. Use mirrors, shiny metal, grease-coated unpolished metal, or improvise a mirror from a food tin or belt buckle. After punching a crosshole in the center of a shiny object, reflect sunlight to a nearby surface, slowly bring it to eye level, and look through sighting hole until a bright spot of light appears on the target. Continue sweeping the horizon even though no ships or planes are visible. Mirror flashes can be seen for miles even on hazy days.

Shadow Signals

If nothing else is available, make signals which cast sun shadows. Shadow signals must be located and arranged to produce the best possible shadow. Near the equator, a north-south line gives a shadow at any time except noon. Further north or south, an east-west line or some point of the compass in between gives best results. Shadow signals should be made in an international distress pattern. Brush, foliage, rocks, or snow blocks piled in a line are effective.

Color Signals

Color attracts attention. White airplanes down on snow or green airplanes in forests are nearly impossible to see. Searchers can only see bright contrasting colors such as yellow-orange, light red, and white against green vegetation. If wearing dark clothes, change to a lighter color. Use white underclothes as flags or change the color with dye or paint. Small signal panels or flags are more visible when moved slowly or waved.

Other Signalling Methods

Repeated sounds coming from unusual places or at odd times attract attention. Whistles, drums, gunshots — anything audible, such as hitting a big stick on an airplane or hollow log, can carry a vital message. If a whistle is not available, make one from a sapling or piece of metal. Yelling or shouting is only effective when a ground party is nearby. If searchers are looking, change surrounding landscape. In brush, cut conspicuous patterns in vegetation; in snow, tramp large side trenches or pile brush or rocks to spell out message — S.O.S. — H.E.L.P. Place trenches or piles so the sun creates shadows to make obvious contrast to the normal scene.

Use available flares, two-way radios, or other signalling devices only when most noticeable. Small radios and emergency beepers have limited range and need to be on high ground for maximum effectiveness.

In the woods, spread colored tarps over tree tops, or hoist a large white or colored flag on a pole lashed to the top of a tall tree.

Do everything possible to disturb the "natural" look. There are very few continuous straight lines in nature and very few right angles. The more that signals contrast with background the better.

At night use flashlights, strobe light, recognition light, or vehicle lights and flashers. Light can be seen several miles over water. Electronic distress markers are extremely effective.

A whistle at night or in a fog can attract surface vessels, people on shore, or locate another separated raft.

The most important factors about signals are:
- Know how signals are to be used. (International Distress Signal, mirror, etc.)
- There are few straight lines and right angles in nature.
- Have signals ready for immediate use.
- Use them in a manner which will not jeopardize safety.
- Use a ratio of six to one on letters used. (Example: 3 feet wide and 18 feet long.)
- Signal with anything and everything possible. It can shorten the emergency and save lives.
- Signals make you effectively larger. Contrast is the key to effective signals.
- Chances are something will be noticed if it is out of the ordinary.

Successful signalling can speed rescue, and eliminate the possibility of a long, uncomfortable ordeal.

Signalling Tips

- For maximum efficiency, place signal fires in open areas. Trees tend to disperse smoke. (Important, use proper precaution in preparing fires.)
- Stay with your stalled or crashed transportation. If you must travel, leave a note and signs of your direction of travel.
- Your aircraft provides enormous supplies and possibilities for signalling.
- *Signalling, like anything else in emergency situations, must be accomplished without further danger and in such a way as to improve your chances for rescue.*

Improvising From the Aircraft

In any emergency, improvise basic needs from whatever natural or manufactured materials are available. It may be necessary to make tools, build shelter, find and purify water, and forage for food.

The greatest survival problem is keeping the body alive a little longer. The heart must pump, the brain must think, and arms and legs must move to reach safety. Money and equipment are useless unless they help keep the body alive long enough to survive the emergency.

Improvising necessities means a person must reason, compare, analyze, and be determined to survive. A person may make a number of mistakes, but with determination, that person will succeed.

Once it is decided something is needed to maintain life, it is imperative to satisfy that need any way possible without violating ethics. Use basic knowledge of physics, chemistry, mechanics, and common sense to utilize what is available to do the needed job. It may not be perfect, but adequate enough for short-time emergency use. As soon as a problem is recognized, decide what is required, then decide how to do it. Be positive, look around and make what is needed from materials at hand. Classify needs and priorities so time is not wasted.

Life depends on maintaining body temperature. Body shelter takes top priority during storms and temperature extremes. Wind, rain, and cold extract heat faster than the body can produce it. Hot sun beating down on hot ground can cause the body to gain heat faster than it can dissipate it. Adequate shelter is vital to survival. Improvise to live.

Attempt to make what is needed from what is available — from items carried or nearby. Look, think, then look again. When lives are at stake, everything is expendable.

Survivors have used chunks of rotten log for building blocks, and bark stripped from fallen trees for floors, walls, and roofs. There are stories of stormbound pioneers who wore bark slab clothing. Nothing is useless in survival. A car, downed airplane — any manmade object — contains a treasure trove of material. Any piece of metal can be sharpened for cutting. Grind it on a rock, chip it, to make it cut better.

If shelter must be built, utilize everything available. If an axe or saw is handy, it is fortunate. You can dig with a hard hat, pry with a straight limb, hammer with a rock, or pull bark off trees with hands.

Thinking, analyzing a need and materials at hand to fulfill it, making it work — all spell improvisation, and could mean the difference between life and death in a survival situation.

> *Articles to improvise from*
> - *What you are wearing, what is in your pockets, other personal equipment*
> - *natural environment*
> - *garbage and rubbish*
> - *stalled transportation*

High Priority Needs

In a survival situation identify the basics for keeping alive. Most immediate is first aid and shelter, clothing materials. Procurement, storage, and retention of water is your next concern. Then concentrate on signalling materials, as well as fire making tools and resources to burn (both for heating and signalling).

Tool Kit

Many kinds of tools are available in combination or in small kits. Look at the aircraft and decide what you might need to dismantle it. This tool kit is imperative to utilize various parts of the aircraft. It should include a combination pliers and cutters, small file and adjustable wrench, screwdriver (preferably with interchangeable heads for standard and Phillips) and a metal saw or heavy duty shears for cutting aircraft skin.

Rules for Improvising

Look at resources and materials as an aborigine would look at them: How can I use this piece of material to provide a basic need? Think of an object that normally does the job. Look at how the object is constructed, materials it is made of, and principles it incorporates. Now look around and see what is available to you. Can you construct it from what you have? Try it. Use failures as incentives to try again. It's done!

Improvised Uses for Aircraft Parts

Fuselage—(Both cockpit and rear of aircraft)—moderate weather shelter.
Wings—Windbreaks, shelter supports, overhead shade, platform for fire on snow, water collection for dew and rain, signal if layed out in a clearing.
Vertical Stabilizer—Shelter support, platform, water filter when inverted.
Aluminum Skin—Reflector for warmth around fire, signal, reflector oven, shade, fire platform, splint material, snow saw blade.
Fabric Skin—(Fire starting material?) Water collection.
Engine Oil and Gas—Fire starter and fuel for stove, signal with black smoke.
Engine Mags—Spark producers for fire starting.
Engine Cowl—Shelter, water collection, windbreak, fire platform.
Nose Spinner Cone—Bucket, stove with sand, oil and fuel, scooping tool in snow, container in solar still, pot for heating water and cooking, funnel.

Propeller—Shovel, snow cutting tool, bracing for shelter.
Wing Struts—Pry bar, splint, shelter brace, flag pole for signalling, crutch.
Landing Lights, Strobes, and Clearance Lights—Signal when used with battery, also if battery is good, to provide light by night.
Fuel Cells—Melt snow on black surface in winter, black smoke in fire, lay out on snow for signal, inside lining of shelter or overhead protection.
Spring Steel Landing Gear—Pry bar, splint brace.
Wooden Wing Struts, Braces and Props in Older Aircraft—Fire starter and fuel.
Doors—Shelter, solar still with windows, shade, windbreak.
Seats—Sleeping cushions, back brace for spinal injury, source of sponge rubber for fire starter and signal material, also insulation and ground pad, padding for compresses, sponge rubber for neck support.
Head Liner and Other Inside Fabric—Water strainer or filter, clothing and added protection, bandage material.
Rugs—Ground pad, insulation, clothing, overhead shade.
Air Charts/Maps—Stuff inside clothing to increase insulation.
Battery—Signalling with lights, fire starting.
Control Cables—Rope, snare wire or binding for shelter, repair cord.
Control Pulleys and Cables—Block and tackle.
Wing Tips—Drip collectors, and water carriers.
Clearance Light Covers—Utensils and tools.
Wheel Faring—Water storage or collection, if broken will produce black smoke in fire.
Tires—Fire starter and fuel for signalling with black smoke, also cut out for carrying container.
Inner Tubes—Cut hole for canteen, cut in strips for elastic binding material, burn for black smoke.
Bungee Cord on Older Planes—Sling material for sling shot, or Hawaiian sling.
Windows—Water collection, solar still, burn for black smoke signal, cut up for snow cutting tool.
Ailerons—Snow cutting tools, shelter braces, splints.
Wiring—Binding, cordage and rope.
Air Filter—Fire starter, improvised water filter.
Oil Filter—Burn for black smoke.
Hoses—Siphoning and also to burn for black smoke.
Fuel Strainer—Standard and Phillips screwdriver.
Landing Light Lense—Fire starting.
Compass—Establishing direction for signalling and also oil for starting fires.
Rotating Beacon Lense—Cup.
Seatbelts—Binding material, slings, bandage.
Battery Box—Stove or container to cook in.
Foam From Seats—Bandaging, insulation, water filter, fire starter, signalling with black smoke.
Brake Fluid and Kerosene From Turn and Bank Indicator—Fire starter.
Magnesium Wheels—For signalling.
Disk Brake Plates—For signalling.

Note: This is only a partial list of possibilities used as an example. There are obviously almost unlimited other uses for each part of the aircraft.

Cold Water Immersion Techniques
Survival Swimming

Most people don't realize that if they are injured and conscious they can survive in all but the coldest and roughest seas. A simple technique called "survival floating" or drown proofing, was developed many years ago. The technique is based on two premises: 1) almost everyone will float while their lungs are full of air; and, 2) it is much easier to float vertically than horizontally.

How to do it: while floating vertically with hands limply held at the sides, take a deep breath and hang relaxed in the water with face below the surface. As the need to breathe arises, exhale through the nose slowly while raising arms and crossing them in front of the face. Then, as if parting a curtain, extend arms and push downward with palms toward the sides and at the same time tilt your head back. As your mouth comes out of the water, take a breath, lower your head and relax again. If relaxed with this technique, alternately switch between slow, horizontal movement toward shore and "survival float." Even with a cramp it is possible to stay afloat. Relax the muscle by slowly massaging it between breaths of air.

> **This technique should only be used when you are without a lifejacket or other improvised flotation.**

No Lifejacket

Three options are open to a person in cold water without a lifejacket. First, is to swim for the nearest refuge or shoreline — distance is the chief factor. If there is no refuge within safe swimming distance, there are two choices. If "survival swimming" or the drown proofing method of conserving energy and staying afloat is used, the primary heat radiating area of the body (the head) is repeatedly immersed. Drown proofing causes body cooling 82 percent faster than a person holding still in a lifejacket. This appears to be the fastest way to bring on the effects of hypothermia in cold water. The third option — slowly treading water — only loses heat 34 percent faster than holding still in a lifejacket. Treading water slowly is definitely the best option to extend survival time.

> **Carry lifejackets and a life raft if you plan on flying over open water for any length of time.**

Immersion Hypothermia

Any pilot or passenger flying over cold water areas should understand the factors that determine body cooling rate while immersed. This vital water safety knowledge, properly used, can extend survival time from minutes to hours. In terms of rescue, this could be the difference between life and death.

Hypothermia is the cooling of the body inner core. In immersion hypothermia, the rate of cooling is greatly accelerated. In water, skin and peripheral areas cool very rapidly because of the increased body surface in direct contact with cold water molecules.

Sea water freezes at 28 to 29 degrees F. and water near ice is usually this cold. Fresh water is usually about 32 degrees F. near ice. Depending on clothing worn, a person falling into this water has the breath knocked out, suffers initial shivering, then goes into a spastic fetal position — hands and knees under chin with no control of voluntary muscles.

In water as cold as mentioned above, body core temperature falls very rapidly. Upon submersion, reflex contractions of smaller blood vessels gives fleeting increase in blood pressure and heart rate. The victim is unconscious in five to seven minutes, and dies in 10 to 20 minutes.

Survival Time

Varying circumstances in every situation will affect individual cooling rates. Body energy levels, type of clothing, metabolic rate, circulatory problems affect survival time. In addition, body fat, physical size, age, and sex influence survival time in cold water. The following table, based on University of Victoria, B.C. research, gives general survival times of average men and women holding still in 50 degree F. ocean water, wearing standard lifejacket and light clothing.

WATER TEMPERATURE DEGREES F.	MAXIMUM TIME OF IMMERSION FOR SURVIVAL
32	1/4 hour or slightly more
36.5	1/2 hour
41	1 hour
50	3 hours
59	7 hours
68	16 hours
77	3 days or more

Should You Swim or Keep Active to Stay Warm?

Exercise increases body temperature and expends calories. This is an accepted way to maintain body temperature in open air, but it is definitely not the case in water. Increased activity in water substantially increases inner core cooling rate due to increased circulation to arms, legs, and skin. The cooling is caused by continuous movement of water over the surface of the body. The cooling effect is similar to convection losses in wind, but much more severe. Research shows the average person swimming in a lifejacket cools 35 percent faster than if merely suspended in water. Consider how far it is safe to swim in cold water. The average person can swim about .85 miles in 50 degree F. water before being incapacitated by hypothermia. As a rule of thumb, if water temperatures are near 50 degrees F., do not try to swim unless within one mile of shore or rescue.

How to Increase Survival Time with a Life Jacket

Infrared pictures show the body's primary heat loss areas are the head, sides of the chest, and groin area. If immersed, concentrate primarily on insulating these areas. Hugging arms close to sides of chest insulates rib cage. Crossing legs and assuming a semi-fetal position with the head out of water can increase survival time up to 50 percent. If more than one person is involved, they should huddle together to conserve energy.

Lifejackets

Most standard lifejackets provide little or no protection from immersion hypothermia. Kapok-filled preservers offer no significant protection. Foam-type vests offer 50 percent to 75 percent increase in predicted survival time and recent research with specifically designed jackets and full suits have increased survival time four fold and more. It is an individual responsibility to decide the type of protection most suitable for personal needs. Adequate cold water protection is available and everyone should acquire it and use it.

Determining Time and Direction

During the Day

In an emergency, it may be necessary to navigate without map and compass. It is necessary to improvise and observe natural phenomena in the general area. You can use the sun to find north (and any other direction, once north has been located) using a branch or stick placed in the ground to cast a shadow.

Using a fairly straight stick about three feet long
- find a fairly level, brush-free spot. Push stick into ground, inclining it to get a longer, bigger shadow if necessary.
- mark tip of shadow with stick, stone, etc. Wait until shadow tip moves a few inches (10 to 15 minutes with a three-foot stick).
- mark position of new shadow tip.
- draw a straight line from first marker through, and about a foot past, the second marker.
- put toe of the left foot at first marker and toe of right foot at the end of the line we just drew.
- we are now facing true north, with left foot at the west end of the line and right foot at the east end of the line.

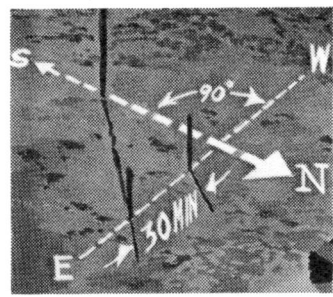

This system works because the sun always travels precisely east to west, even though it might not rise and set at exactly 90 degrees and 270 degrees. The shadow tip moves in the opposite direction, so the first shadow tip mark is *always* west of the second, anywhere on earth.

An alternate method also uses a three-foot stick, this time pushed into the ground so it points directly at the sun, casting no shadow on the ground at the base of the stick. Wait 10 or 15 minutes and check the direction the shadow falls. A line drawn through the tip of the shadow and the base of stick is an east-west line. The narrower the stick is at the top, the more accurate direction will be.

The shadow tip method used for finding direction can also determine approximate time of day.

- After drawing an east-west line, drawn an intersecting, perpendicular north-south line. Push a stick into the ground at the intersection. The west part of the line indicates position of the stick's shadow at sunrise (arbitrarily 0600 hours). The east part of the line indicates shadow position at sunset (1800 hours). The north-south line indicates shadow position at noon.
- The shadow of the stick becomes the hour hand of your clock and you can estimate time using the noon line and 6 o'clock lines as guides.
- The shadow clock is not as exact as a watch. It divides the day into 12 unequal hours, with sunrise always 0600 and sunset always 1800.
- Twelve o'clock shadow time is always true midday, but spacing of other hours varies somewhat with location and time of year.

An ordinary wrist or pocket watch can help find true north. In the north temperate zone (from latitude 23-½ to 66-½ degrees) point the hour hand toward the sun. A line halfway between the hour hand and 12 o'clock (standard time) or halfway between hour hand and one o'clock (daylight time) is a north-south line. On cloudy days, hold a small stick at the center of the watch so its shadow falls along the hour hand. A line halfway between the shadow and 12 o'clock points north.

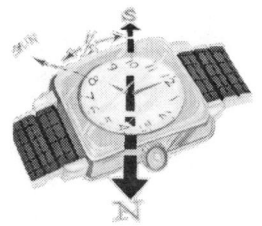

At Night

To find the North Star, locate the Big Dipper. The two stars forming the end of the bowl are "pointers". Visualize a straight line drawn through the pointers. Along this line, about five times the distance between the pointers, is the North Star — above the Dipper lip. The Big Dipper rotates around the North Star, and is not always at the same point. When the Dipper is low in the sky, and may be obscured by trees or high ground, the constellation Cassiopeia, a group of five bright stars shaped like a lopsided M (or W when low in the sky) can also be used. The North Star is straight out from the middle star in Cassiopeia, at about the same distance as from the Dipper lip. Since Cassiopeia is almost directly opposite the Dipper, one constellation will generally be observable for finding the North Star.

Direction From Orion

The constellation of Orion consists of seven stars. The three close together are called the Belt of Orion. The star through which the north-south line on the diagram passes is exactly on the Celestial Equator. No matter where on earth you are, this star rises due east of you and sets due west.

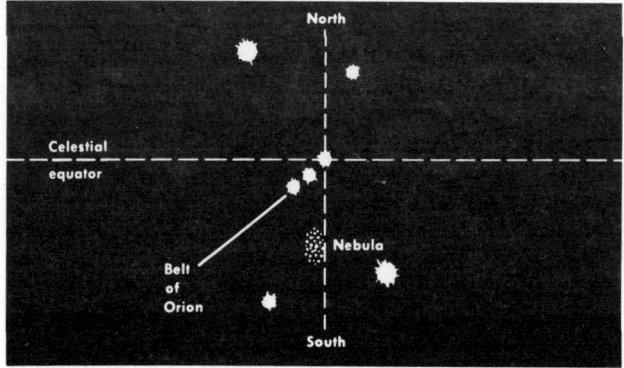

Determining Time with the Big Dipper

Time at night can be determined by using the Big Dipper. Consider your location within a time zone, and whether it is standard or daylight time.
- Visualize a standard clock around the North Star.
- Facing directly north, determine what hour the Dipper pointer stars point closest to. Subtract this number from 12.
- Multiply by 2.
- Add 11 to determine local star time, or 24-hour clock time.
- Subtract two hours for each full month since March 23, and another four minutes for each remaining day. The result is hours and minutes in conventional time.

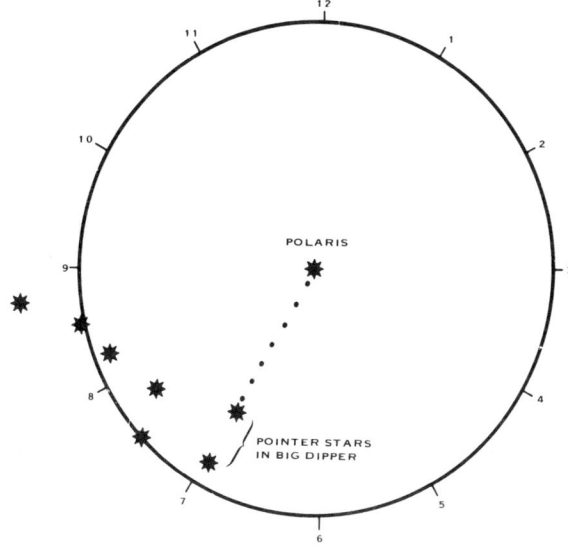

EXAMPLE: August 20th, 7 o'clock on the star clock:
1. 7 from 12 = 5
2. 5 X 2 = 10
3. 10 + 11 = 21 (2100 hours on the 24-hour clock)
4. March 23 to August 20 = 4 months, 27 days.
   4 months at 2 hours/month = 8 hours
   27 days at 4 minutes/day = 1 hour 48 minutes
   2100 hours minus 9 hours 48 minutes = 11:12 P.M.

Determining Time with Cassiopeia

Time can be determined by using the star Caph, in the constellation Cassiopeia and the diagram in the figure on the next page. The 12 months are printed on the outside of the circle (all months are considered to be 30 days), and 24-hour clock hours are printed on the inner circle, reading counterclockwise.

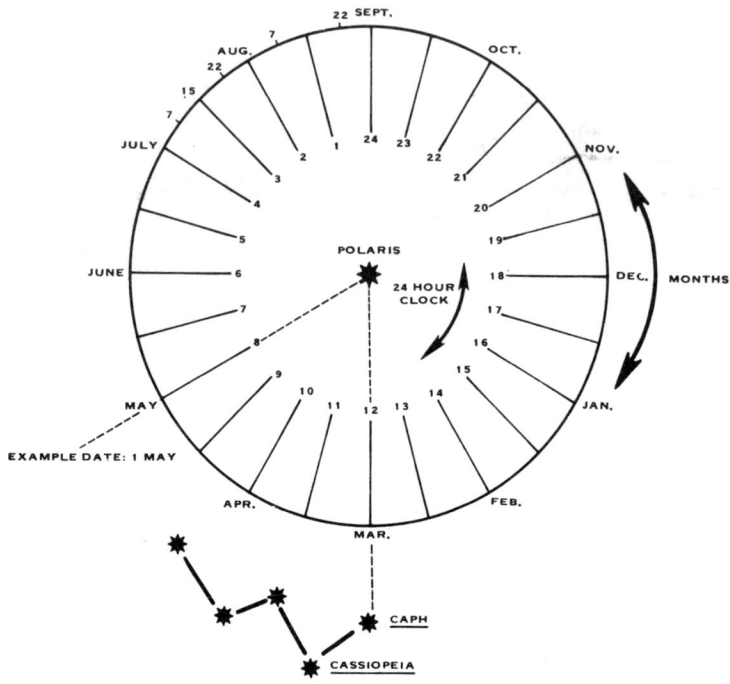

To determine time on May 1st:
1. Read 24-hour clock for May 1.
   Place number here — 0800
2. Read 24-hour clock for Caph.
   Place number here — 1200
3. Add the two numbers.
   —2000

This then indicates the time on the 24-hour clock. (When total exceeds 24, subtract 24 for correct time on 24-hour clock.)

Water Procurement and Treatment

The human body is approximately 75 percent water. The daily intake and output of liquids and certain chemicals are necessary for life processes and normal functions of vital organs. The average daily body water rquirements for proper biological balance and efficiency is two quarts of water per day. Any lower intake results in gradual dehydration and the loss of mental and physical proficiency. As one of the major "necessities of life", body water loss and the accompanying problems of thirst and hypohydration will be important considerations during any emergency situation.

Humans lose body water in three ways: through perspiration, respiration, and through the elimination of solid and liquid waste. The first two involve body processes that remove excess heat from the body. The waste products of food oxidation and muscle or energy use are removed by bowel movements and urination. Though it is possible to minimize body water loss, you will still lose water on a daily basis through these automatic body processes. Losing water to the extent of 2.5 percent of body weight (about 1½ quarts) will reduce your mental and physical efficiency by 25 percent. Your sweat rate while walking in hot environments can increase to two quarts per hour.

Signs and Symptoms of Dehydration at Various Stages of Body Water Deficiency

1-5% of Body Weight	6-10% of Body Weight	11-20% of Body Weight
Thirst	Dizziness	Delirium
Vague discomfort	Headache	Spasticity
Economy of movement	Labored breathing	Swollen tongue
No appetite	Tingling in limbs	Inability to Swallow
Flushed skin	Decreased blood volume	Dim vision
Sleepiness	Increased blood concentration	Shriveled skin
Increased pulse	Absence of salivation	Painful urination
Increased rectal temperature	Cyanosis (blue body)	Numb skin
Nausea	Indistinct speech	
	Inability to walk	

Water requirements are critical in any emergency environment. In the desert, your life will depend upon your water supply and how well you manage and conserve your body water. The table on the next page will give you an idea as to how long an average person could be expected to live on certain amounts of water at various temperatures. The minimum daily requirements of two quarts of water per day increases to a gallon or more in hot environments.

Generally search for water and build shelter near it. If none is near improvise water catch basins for rain and dew. Look for green vegetation or moss, and water holding plans such as the barrel cactus. Obtaining water may not be easy even if wet ground is found. You can wring water out of vegetation, use stills, or dig catch basins for seepage water.

Days of Expected Survival in the Desert Under Two Conditions

©Physiology of Man in the Desert, E.F. Adolph and Associates, New York-London Interscience Publishers, 1947.

NO WALKING AT ALL

MAXIMUM DAILY TEMPERATURE (°F) IN SHADE	AVAILABLE WATER PER MAN, U.S. QUARTS					
	0	1 Qt	2 Qts	4 Qts	10 Qts	20 Qts
	DAYS OF EXPECTED SURVIVAL					
120°	2	2	2	2.5	3	4.5
110	3	3	3.5	4	5	7
100	5	5.5	6	7	9.5	13.5
90	7	8	9	10.5	15	23
80	9	10	11	13	19	29
70	10	11	12	14	20.5	32
60	10	11	12	14	21	32
50	10	11	12	14.5	21	32

WALKING AT NIGHT UNTIL EXHAUSTED AND RESTING THEREAFTER

MAXIMUM DAILY TEMPERATURE (°F) IN SHADE	AVAILABLE WATER PER MAN, U.S. QUARTS					
	0	1 Qt	2 Qts	4 Qts	10 Qts	20 Qts
	DAYS OF EXPECTED SURVIVAL					
120°	1	2	2	2.5	3	
110	2	2	2.5	3	3.5	
100	3	3.5	3.5	4.5	5.5	
90	5	5.5	5.5	6.5	8	
80	7	7.5	8	9.5	11.5	
70	7.5	8	9	10.5	13.5	
60	8	8.5	9	11	14	
50	8	8.5	9	11	14	

Along the Seashore

Along coasts, fresh water may be found in dunes above the beach, or in the beach itself well above the high tide line. Check hollows between dunes for visible water. Dig if the sand is moist.

How to Dig
- Along the seashore on sandy beaches, dig a hole just over the first sand dune that is directly adjacent to the water. A damp area, in a depression, is a good place to dig. This is where rainwater and drainage from the local water tables collect.
- Dig until the hole begins to fill with muddy water, a deep hole is not necessary. (If you dig too deep, you may reach salt water, which is unfit to drink.) Shore up the sides of the hole with beachwood or brush.
- Let the impurities and suspended particles settle and clarify the water. Make the water safe by purification and clarification.

> **Do not ever drink sea water or urine. The salt content is too high.**

In Desert or Arid Lands

Analyze the emergency situation and water requirements. What is the probability of being reported overdue? *Did you file a flight plan?* If you expect to be missed and can expect a search to begin within 12 to 24 hours, then it may be wiser to sit and conserve body water while awaiting rescue.

> **All efforts to find water by search and digging will consume energy and water, and will increase your water requirements.**

Do not rely on the sensation of thirst to indicate how much water you need. Drink plenty of water any time it is available, particularly when you are eating. The main way to conserve your water is to *control your sweating.*

How to Control Sweating
- Keep your clothes on. Clothing helps control sweating by not letting perspiration evaporate so fast. Clothing also prevents sunburn.
- Wear a hat, use a neckcloth.
- Light-colored clothing reflects light and heat.
- Keep in the shade during the day.

> **Ration your sweat, not your water.**

Where to Look for Water in Desert and Arid Areas

Water is more abundant and easier to find in loose sediment than in rocks. Look for springs along valley floors. Flat benches or terraces of land above river valleys may have springs or seepages along their bases, even when the stream is dry. Signs of damp sand along the bottom of a canyon or base of a hill can be dug in as described in the previous seashore section. Dry stream beds may have water just below the surface. Try digging at the lowest point on the outside of a bend in the stream bed channel.

Animal trail forks usually point toward a source of water. As the animals move from their various habitats, they converge more and more to the same trails as they approach water. Watch for animals and birds moving in the early morning or late evening. They are probably moving toward water.

Vegetation Water Still

A new innovation in solar stills has been recently introduced which appears to be superior to the traditional desert solar still. Current tests are investigating possible contamination of water if poisonous plants are used. Water produced tastes much like the plants used, but the idea is sound when poisonous plants are avoided.

How to Build a Vegetation Water Still
- Scoop out a double craterlike hole on a slope or built-up mound of dirt.
- Place the plastic bag over the crater, with the bag opening down slope.
- Inside the bag, use a stick or something similar to make a tent over the crater. Pad the ends of the stick.
- Place clean rocks or weights inside the bag around the outer crater rim to hold the plastic tent taut.
- Fill the inner crater with vegetation. Use care not to have any vegetation touching the outside crater or the plastic.
- *Close the bag tight.*
- As the vegetation warms up, water will condense on the inside of the plastic and droplets will run down the tent and collect in the outer crater. (Tests have shown that the vegetation still works faster and produces more water than the traditional desert still.)
- From the outer crater, water runs down the trough to a catch hole or pool.
- To get water from the bag, scoop a deeper hole just below the catch pool and open the bag just enough to pour out the water.

- Caution—Avoid water that was in direct contact with the vegetation. Do not drink the water formed under the vegetation. Some garbage bags are treated with anti-fungus or bacteria chemicals. Do not use.

    Note: Colored plastic bags will work, but clear plastic is the most efficient.

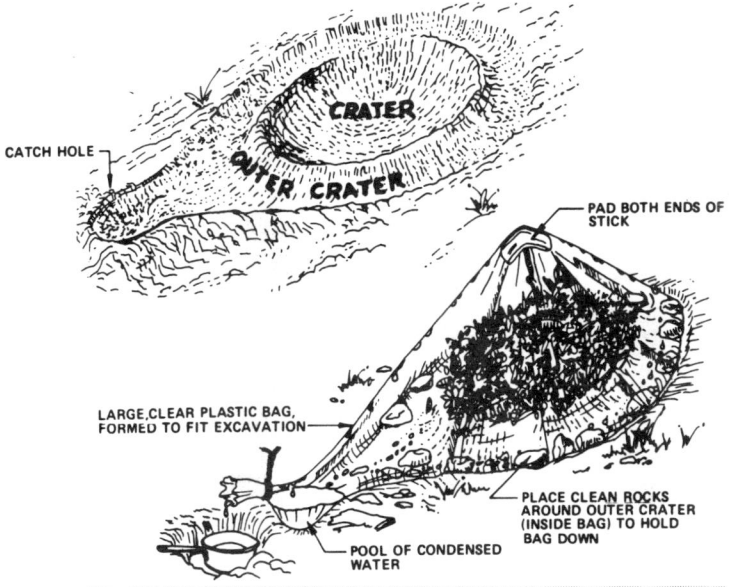

Garbage bags (not treated chemically) have many uses. Your emergency preparedness kit should contain several.

Desert Solar Water Still—Conventional

Solar stills work slowly and require hot sunlight and patience. When properly set up and lined with vegetation, they can produce a pint of water in about three hours.

How to Build a Desert Still
- Dig a hole three feet wide and three feet deep in the lowest possible location where water would stand the longest after a rain.
- Line the bottom of the hole with any green vegetation available.
- Place a water container in the bottom, at the center.
- Install a water sucking tube, if available.
- Place clear plastic across the hole so that it drapes down in the center above the container.

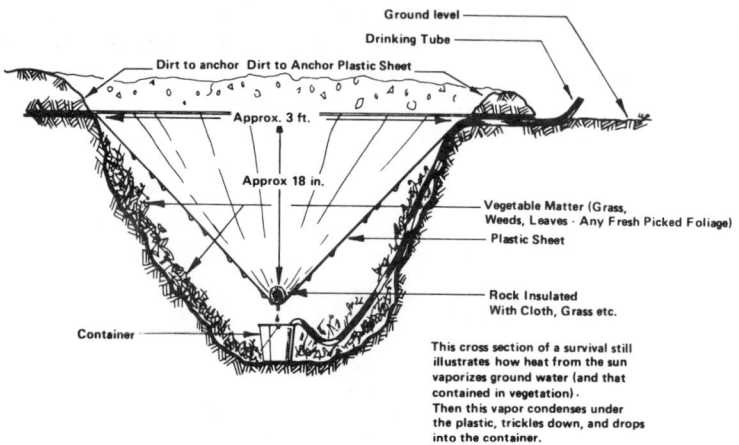

This cross section of a survival still illustrates how heat from the sun vaporizes ground water (and that contained in vegetation). Then this vapor condenses under the plastic, trickles down, and drops into the container.

- Seal the edges of the plastic with dirt.
- Place a small rock, insulated with cloth or paper, in the center of the plastic, directly over the container.
- Note: *Clear plastic is best. Semi-transparent colored plastic may work. Dark colored plastic blocks out the sun's rays and may not work.*

A still should be opened only when necessary because it takes a long time to regain the operating atmosphere. Use the drinking tube to suck out the water in the container. Set the still up in the early morning while it is still cool. Let it work while your rest in your shade shelter. Make several stills if you have the materials.

> *If you are planning to fly over arid areas, your emergency preparedness kit should contain the materials necessary to construct solar and vegetation stills. Water is critical to human existence. In the desert, life will depend on water supply and how well body water is managed and conserved.*

Water in the Mountains

A mountainous area is often the easiest geographic environment in which to locate water. The amount of difficulty will depend upon elevation, vegetation, geologic structure of the mountains, and annual precipitation.

Most canyons have streams, springs, or intermittent run-off flow during all or part of the year. Look in the very bottom of the canyon floor near catch pools and depressions of stream beds. If there is water visible, often a small amount of digging will produce water.

Where to Look for Water
- Look for large rock formations with green moss or lush vegetation and seek water at the base.
- Small clumps of isolated green vegetation in arid mountains is a good sign of a spring.
- Collect dew from leaves, plastic sheets, metal of the airplane, or any wet surface. Sponge the water off the absorbent cloth.
- Improvise rain-water collectors. Catch rain in plastic sheets or waterproof clothing.
- *Caution: Waterholes with few or no tracks and plants nearby may have a high mineral content which could cause extreme illness.*

Arctic and Winter Conditions

Water procurement and hypohydration is almost as great a problem in cold conditions as in the desert, because all the water is frozen into snow or ice.

Where to Look for Water
- Streams or lakes may provide access to water.
- If the sun is shining, you can melt snow on dark plastic, a dark tarp, metal surfaces of the airplane, or any surface that will absorb the sun's heat. Arrange the surface so the melted water will drain into a hollow or container.
- Whenever possible, melt ice for water rather than snow. You get more water for the volume with less heat and time.
- Note: If planning on flying over arctic or winter environments, your emergency preparedness kit should contain a stove and pot to melt ice and snow.

Do not eat ice or snow. A day or two of eating ice or snow results in swollen, raw mucous membranes in the mouth. Also, the expenditure of body heat and energy to melt the snow or ice is too costly.

> **Never underestimate the importance of water during any activity.**

Preparation of Water Under Emergency Conditions

Symptoms of water illness are nausea, vomiting, diarrhea, low grade fever, vague feeling of discomfort, fatigue, and weight loss. It might be necessary at some time to utilize muddy, stagnant, or polluted water. This type of water may not seem palatable, but if treated properly it will be harmless and will satisfy your need for water.

Clarification

Clarify muddy water before purifying. You can get water that is almost clear from muddy streams or lakes by digging a hole in sandy soil one to six feet from the bank. Allow water to seep in, then wait for it to settle. Improvise a filter out of a can or cloth (sock, sleeve, etc.) using layers of sand at the bottom, then a layer of charcoal, then a layer of grass on top. Multiple layers of cloth make good filters. If you have containers, let muddy or turbid water stand for a day. Be gentle when pouring clear water off of the sediment.

> **Remember, clarification is not purification. Water that is clear and pure in appearance can be highly contaminated with bacteria.**

Purification

The safest method of purifying water is to boil it vigorously for one minute plus an additional minute for each thousand feet of altitude. If in doubt of altitude, boil for five minutes. Iodine purification is another method. Using 2 percent tincture of iodine, add four drops to each quart of clear water, eight drops to each quart of cloudy water. For a gallon, add 12 drops for clear water, 24 drops for cloudy water. Stir thoroughly.

For water purification tablets follow directions on the label. Usually one tablet is sufficient for one quart of water. Use two tablets per quart if the water is cloudy. For bleach purification follow the chart below. It shows the amount of bleach to add, using bleach solutions containing 5.25 percent of sodium hypochlorite. Add the chlorine solution to the water and stir. Let stand for 30 minutes. After this length of time, the water should have a distinct taste or smell of chlorine. If not, add several more drops of chlorine, and let stand 15 minutes. The taste or smell of chlorine in the water is a sign of safety.

Amount of water	Amount of solution to add to—	
	Clear water	Cloudy water
1 quart (¼ gallon).	2 drops	4 drops
1 gallon————	8 drops	16 drops
5 gallons————	½ teaspoon	1 teaspoon

Food Procurement and Preparation

In considering immediate survival priorities, food has previously been placed last in order of importance. However, an emergency extending beyond three to four days takes on aspects of long-term survival. Assuming more immediate priorities have been met, resupplying body energy now moves into proper perspective with regard to priority. People have gone amazingly long times without caloric intake; however, when body blood sugar level drops below a certain point, possible detrimental reactions occur.

Practically all body cells can live on fatty acids and protein constituents as well as sugar. Brain cells, however, live exclusively on glucose and a sharp drop in blood sugar poses immediate threat to the brain. To guard against this danger, the hypothalamus continually monitors blood sugar. If for any reason — such as stress or strenuous exercise — the renewal rate is inadequate, the hypothalamus stimulates the adrenal gland releasing a large charge of adrenalin into the bloodstream. When adrenalin reaches the liver, it stimulates breakdown of glycogen into glucose for the brain's increased demand.

Elsewhere, adrenalin conserves sugar by preventing many body cells from feeding on the dwindling stores. The body will go to almost any lengths to maintain the glucose supply needed by the brain. It will drain itself of fat, even consuming muscle to the point of weakness and wasting away to avert damage the brain cannot withstand. As the emergency continues, coordination, dexterity, and endurance begin to drop off rapidly. Depending upon demands of the situation, the resupplying of body energy level becomes increasingly important as days pass. If an emergency lasts five days or more, you may have to resupply the body's fuel tank with food procured from the immediate area.

Some important physical effects of undernutrition and starvation are expressed by the following:
- Hunger and hunger pangs
- Weight loss
- Weakness, decreased muscular endurance and strength, poor coordination and halting movements
- Dizziness and blackouts when standing up suddenly
- Slowed heart rate
- Increased sensitivity to cold
- Increased thirst accompanied by a craving for salt

Physical reactions to starvation show body effort to adapt and protect itself against hunger stress. Decreased heat rate indicates functioning at a lower energy level, thus conserving energy resources. Objective tests show vision stays sharp and hearing actually becomes more acute.

Constant hunger pangs, coldness, weakness and loss of endurance are not conducive to a positive attitude. Psychological effects include personality and behavior changes, depression, irritability, nervousness, and general emotional instability; social withdrawal, decrease of communication and social interchange; narrowing of interests with emphasis on thoughts of food; and, difficulty in concentration and slow speech.

These physical and psychological effects are reversed by food and a protective environment. Food contributes to morale and replaces substances burned to provide energy for survival. Controlling hunger is quite easy if one can adjust to discomfort and adapt to primitive conditions.

If well-motivated and previously well-nourished, a person can survive a relatively long time using resources stored within the body.

It is possible to find something to eat in almost every location. Simple techniques described in this book are by no means complete. Only very basics of how to find and prepare natural foods using materials at hand are discussed. More detailed information is available from comprehensive long-term survival manuals. Even though we may not need food for physiological survival, knowing that we can find and prepare it contributes to confidence and a positive mental attitude.

Procuring Animals for Food

Where can food be found and what is looked for in determining what is edible? Consider animals as a possible source.

Animal food provides the most food value per pound. Anything that creeps, crawls, swims, or flies is a possible source of food. People eat grasshoppers, hairless caterpillars, wood-boring beetle larvae and pupae, ant eggs, and termites. Such insects are high in fat. Many often eat insects as contaminants in flour, corn meal, rice, beans, fruits, and greens of daily food.

Most animals are wary and hard to catch. Hunting them requires skill and patience, and energy expended in long hours of hunting may far exceed return on usable energy to the body's fuel tank. The best method of procuring animals is simple traps and snares. After initial effort of setting out snares, they hunt for a person 24 hours a day.

Time can be spent on other important functions to improve the situation. Getting food with simple traps or snares does not require extensive training.

Finding and Obtaining Animals

Areas which produce most success with trapping and snaring must be located. Knowing something about animals and their habits is useful in placing snares or traps.

Game signs may be found in the following areas:
- Along banks of rivers and streams
- Areas around waterholes, pond, or lakes
- Margins of forests as well as natural meadows
- Marshes, mud flats, and swamps
- Protected mountain slopes, small gulleys, canyons, and draws.
- Areas between high and low tides on the beach
- Abandoned cultivated fields or farm areas

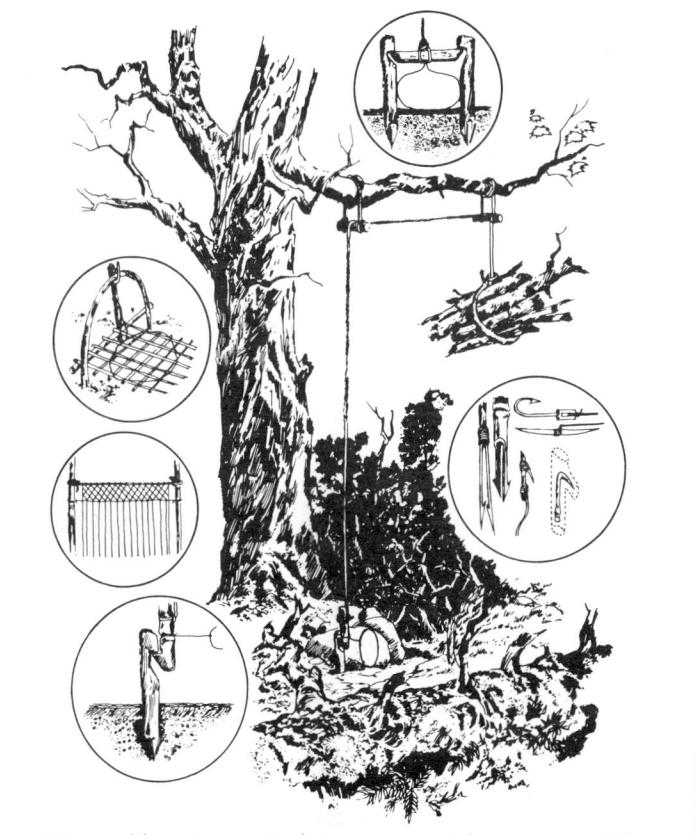

Pictured here is a typical trap set up with counterweight, trigger and snare wire loop. Three other triggers utilizing counterweight and snare wire loop are also pictured. A gill net and other fish procuring devices will also provide food with a minimum of effort.

Poorest areas for finding game are high mountain tops, dry ridges, and dense, continuous forests.

Trails, tracks, droppings, feathers, rubs, hair, beds, and nests indicate presence of game as well as number and size of animals present. Like humans, most animals are creatures of habit and their movements can be anticipated accurately. Feeding habits are carried on from day to day on the same trails and runs used to reach water supply. A basic understanding of the animal desired; i.e., basic habitat, food, and times of movement is needed.

Animals rely heavily on senses to alert them to danger. It is essential that traps or snares be camouflaged for scent as well as sight. Human scent is an immediate danger sign to many animals large or small and must be removed or covered up. Urine from another animal, musk oil from the scent glands of animals, spoiled food, or any strong natural scent can camouflage human odor. Newly-cut wood should be rubbed with soil or some natural coloring agent and bark should be left on carved triggering devices. Plants near the trap should be disturbed as little as possible, and nothing left grossly out of place. (Boughs from trees should not be used on the ground, etc.)

An ideal trap or snare location is one that naturally confines the animal's passageway. The trail or path may pass between two trees, go between obstacles such as logs or rocks, or pass through a particularly thick concentration of vegetation. Further narrow the animal's passage so there will be no last moment to detour to miss the trigger. Funnel the animal through the snare or trap. Use the same principles to set a baited trap for any bird or animal although birds are not usually as wary. Restrict the approach and leave no alternatives for the animals to take.

Trap friction points must not bind. Free and easy travel of counterweight lines is imperative whenever they cross objects for change in direction. All surfaces which touch in a trigger mechanism must be as friction-free as possible. Make a hair-touch release on all sets. Trap counterweights must outweigh the intended animal by at least four times. Counterweight travel must exceed the animal's total length by at least four times to eliminate possibility of reaching the ground or other objects.

Many variations of traps and triggered snares can be used to procure a variety of animals. This book does not contain detailed description of techniques involved in setting traps and snares. There are many such books on the market. Major concepts described here point out principles involved which are not complicated. *(These techniques are strictly illegal in all but survival situation.)*

Preparing Large and Small Animals for Food
- Skin large game and gut all animals. You can drink small amounts of blood.
- Carefully remove gall and urine bladders. If broken, the meat will be tainted.
- Wash meat but not extensively as this makes it tough, stringy, and tasteless.
- Avoid cutting onto or squeezing musk glands. If necessary to remove from skin, use great care. Badgers, civets, skunks, minks, and otter have anal glands. Deer have musk glands on legs; musk deer have them in belly skin. Most of the objectionable odor of small rodents is removed by skinning.
- Bleed and gut birds; skin the larger ones. Carrion-eating birds, such as vultures, have flesh unpleasant to the taste. Fish-eating birds have strong, fish-oil flavor.
- The best meat on a lizard is hind quarters and tail. Legs of a frog are edible. Turtles have edible flesh on legs, neck, tail, and other parts of the body.
- Skin frogs and snakes. Remove and discard skin, head, and viscera.
- Clams, oysters, mussels, crabs, and lobsters left in clean water overnight clean themselves.
- Immediately after landing, a fish should be bled by cutting out gills and large blood vessels next to backbone, then scaled and washed in clear water.

Procuring Plants for Food

Learning what plants to eat and what to avoid is too complex for this text. Interested persons should review the many detailed books on the subject.

There are many easily recognized, easy-to-prepare plants which can provide nourishment during an emergency. Knowledge and ability to find and prepare these foods greatly enhances personal self-confidence and contributes substantially to well-being in survival. The following material deals with generalizations concerning plants.

Of at least the 300,000 different kinds of wild plants in the world, a large number are potentially edible (some more palatable than others). Under survival conditions, wild plant food available may require alteration in diet. Plants are more common than animals so should be used to the fullest.
representative plants and respective uses.
Plant Edibility Rules
- Never eat large quantities of strange plant food without first testing it.

- Take a teaspoonful of the plant, prepared in the way it will be used (raw, boiled, baked, etc.), hold it in the mouth for about five minutes. If no burning sensation or other unpleasant effect, swallow it. Wait eight hours.
- If no ill effects (nausea, cramps, or diarrhea) result, eat a handful and wait eight hours.
- If no ill effects occur at the end of this time, the plant may be considered edible. (New or strange food should be eaten with restraint until you become used to it.)

A disagreeable taste in food which is otherwise safe to eat may sometimes be removed by leaching; i.e., pouring cold or hot water through chopped, crushed, or ground material. Boiling in one or more changes of water may remove the unpleasant taste.

Olives are bitter and grapefruit is sour, so an unpleasant taste does not, in itself, mean poison. But a burning, nauseating, or bitter taste warns of danger. A small quantity of even poisonous food is not likely to prove fatal or even dangerous, whereas a large quantity may be. (This statement does not apply to mushrooms which are best avoided.)

It is not always safe to try foods that are eaten by birds and mammals. Food eaten by rodents (mice, rats, rabbits, beavers, squirrels, muskrats) or bears, raccoons, and various other omnivorous animals are usually safe to try.

Poisonous Plants

As a rule, poisonous plants are not a serious hazard unless a person accidentally walks into a patch of them. The chances of eating a poisonous plant are rare. Frequently, only seeds are poisonous, but care must be used in selecting any plant part. Water hemlock is a violently poisonous plant of the northern temperate zone, and usually occurs in marshy places. This plant belongs to the parsely, carrot, and parsnip family, which contains many well-known edible plants. Avoid all members of this family since water hemlock is fairly common in the Northwest.

Cautions

Cook plant foods when in doubt about edibility. Some poisons may be removed by cooking. Do not eat untested plants with milky juice or let milk contact skin. Avoid eating plants that taste disagreeable (bitterness is a guide).

Avoid ergot poisoning from infected heads of cereals or grasses; discard grain heads having black spurs in place of normal seed grains.

How to Select Edible Plants

Water or land plants furnish edible fruits, seeds, bark, tubers, buds, leaves, flowers, sap, pods, nuts, stems, rootstalks, shoots, and bulbs.

Usefulness of Plants

Common wild plants that we see everyday constitute a vast food supply and useful resource which is often times overlooked. Just because it is not found in a supermarket or store does not mean a plant can't be useful in a variety of ways. Remember to use caution when considering an unknown plant for a food source.

As our domestic species of plants have seasons, so do the wild varieties which grow near by. A few examples are illustrated below.

CATTAIL

Pollen — Gathered in the spring to make bread or gruel.

Seed Pods — Late summer to late fall for fire tinder, improvised torches, insulation and bedding.

Leaves — Summer to fall for clothing and bedding.

Stems — Spring to early summer as a vegetable like young cucumbers.

Shoots — Gathered in the spring and cooked as a vegetable much like asparagus.

Rhyzomes — Spring to fall — peeled and dried, pounded to make flour (very high in starch content).

GRASSES

Seeds — Gathered in late summer to fall, ground and used much the same as wheat.

NETTLES

Leaves — Gathered in spring and cooked just as spinach.

Stalks — Late fall to winter for fibers to be made into twine and cordage.

BARK

Cambium Layer — Survival food all times of the year.

All parts of some plants are edible, but it is usually necessary to select the most palatable root or fruit, leaves or pods. In some instances only the nuts are edible.

Starchy Foods

Many plants store large quantities of edible starch in underground parts. Two examples are the tubers of the cattail (found in most of America) and rootstalks of the Camus, Sego Lily and wild onion.

Bulbs are commonly produced by members of the lily family, such as true lily, onion, tulip, and daffodil. Many bulbs are edible. Tubers, rootstalks, and bulbs are a fine source of food because they are usually available throughout the year. In cold climates, these underground storage organs can be found by digging under dried plant stalks.

Grains or seeds of millet, wild grasses, wild rye, and other grasses, are starchy and excellent staple food. Preparation: cook starchy food as it is difficult to digest.

Mushrooms and Other Fungi

Poisonous fungi cannot be detected by unpleasant taste or disagreeable odor. Although some mushrooms and other fungi are edible, they contribute little food value and are easily confused with poisonous types thus should be disregarded as food sources.

Ferns

Three widely distributed types of ferns illustrate this group of good plants:

1. the bracken
2. the polypody
3. the tree fern

Many ferns are edible; none are poisonous, but they must not be confused with water hemlock. Edible kinds occur mainly in forested areas. Some are only a few inches high. Desert ferns are usually small and tough. In the far north, ferns are few and very small; look for edible kinds in moist, shady places.

The fiddleheads of all ferns are curled, young, succulent fronds which are as good as asparagus in food value with similar taste. Most fiddleheads are covered with hair which makes them bitter, but hair can be removed by rubbing in water.

Preparation: especially bitter fiddleheads can be boiled for 10 minutes and reboiled in fresh water for 30 to 40 minutes. Wild bird seeds or meat can be cooked with fiddleheads.

Nuts

Edible nuts are the most sustaining of all raw forest foods. They are found throughout the world. Many American nut trees such as oak, hickory, hazelnut, and beechnut, are widely distributed through the Northwest states.

Several temperate evergreen trees, especially pines, produce edible nuts. Shake or break open cones to remove edible seeds.

Bark

Inner bark from numerous trees (raw or cooked) may be eaten. In famine areas, flour is made from inner bark of trees. Thin, green, outer bark, and white, innermost bark are normally used. Brown bark ordinarily contains bitter tannin.

Among trees with bark used as food are poplars (including cottonwoods and aspens), birches, and willows. Inner bark and growing tips of a few species of pine including Scotch pine are edible. Pine bark is especially valuable for vitamin C. Outer pine bark is scraped away and inner bark stripped from the trunk to be eaten fresh, dried, or cooked. It may be pulverized into flour.

Bark is most palatable when newly formed in the spring. It is most useful in northern region where food may be scarce, especially in winter. An infusion of evergreen needles can be boiled as tea.

Grasses

Grasses may be the most important single source of survival food, *especially in warmer parts of the world.* Rice, millet, sorghum, maize, and several other cereals are extensively cultivated in the tropics, with rye, wheat, and oats grown here in the temperate regions. Wild grasses have an abundance of seeds which may be eaten boiled or roasted after the chaff is rubbed from seeds.

Preparation: no known grass is poisonous. If kernels are still soft and free of large stiff barbs, they can be boiled for porridge.

To gather grass seeds, place a cloth on the ground and beat grass heads with sticks (winnowing). Many grasses pop like popcorn when heated in a closed vessel. A paper sack can also be used for harvesting by placing grass heads inside the bag and vigorously shaking to dislodge tiny seeds.

Water Plants

Plants growing in wet places along margins of rivers, lakes, and ponds, and directly in water are potential survival food. Succulent underground parts and stems are most frequently eaten. Poisonous water plants are rare. In temperate climates water hemlock is the most poisonous plant around marshes and ponds.

Two kinds of marsh and water plants are the cattail and water lily. Cattail is found worldwide except in tundra forested regions. They also can be found in most moist places in desert regions.

The young shoots which taste like asparagus, and rootstalks, without outer covering, are eaten boiled or raw. While the plant is in flower, the abundant yellow pollen may be mixed with water into small cakes and steamed as bread.

Water lilies occur throughout the world. Temperate water lilies produce enormous rootstalks and yellow or white flowers that float on water surface; rootstalks or tubers may be difficult to obtain because of deep water, but are starchy and full of food. They can be eaten raw or boiled.

Stems can be cooked in a stew. Young seed pods may be sliced and eaten as a vegetable. The seeds may be bitter, but are nourishing. They may be parched and rubbed between stones as flour.

Seaweed

Many seaweeds are edible, but large amounts could be detrimental because they become violent purgatives, although none are actually poisonous. They may be eaten as flavoring in other foods, are rich in iodine, minerals, and vitamins, and will prevent scurvy. Some have too much lime carbonate or are too horny to be eaten. Others are covered with slime. In selecting seaweed for food, choose only plants attached to rocks or floating free. Do not take plants stranded on the beach. A course, dark green seaweed with large air bladders is rockweed. It has no food value, but in and under parts of it are small crabs, shrimps, and shellfish. This seaweed is excellent for wrapping shellfish; it stays moist and keeps shellfish fresh.

Cooking

Cooking without utensils is not difficult. Common sense use of basic principles will make most food palatable. Cooking methods which preserve natural juices and vitamins are best and should be used if possible.

Cooking makes a meal more enjoyable. All wild game, freshwater fish, clams, mussels, snails, and crawfish must be thoroughly cooked for safety. Tough mussels or large snails should be minced. Raw or smoked freshwater fish are frequently contaminated with tapeworm and lung fluke parasites which are destroyed by thorough cooking. Hawks and crows can be tough but soften up when stewed. Plant foods are made more digestible and palatable and yield more food value after heating.

Saltwater shellfish may be eaten raw but are safest cooked. If one or two of the below methods are remembered and used, it can substantially improve morale and comfort.

Some methods of cooking without utensils are the following:
- Roast or fry food on rocks with reflected heat or with rock over flame for improvised griddle.
- Roast foods by wrapping in leaves and burying under a fire in sand or soft soil. Line a small, shallow pit with leaves, then fill with food, or wrap food in plant leaves or cloth. Cover pit

with a quarter to a half-inch layer of sand or soil and build fire directly over it. After sufficient cooking, rake fire away and remove food.
- Roll foods in clay or mud and roast in hot coals. Coat fish, potatoes, shellfish and other foods with a layer of mud or clay. Place directly in flames or coals. Food loss by burning is reduced substantially. One need not scale fish or rough-skinned plants when prepared this way. Peel off skin with baked mud or clay after cooking.
- Bake foods in a reflector oven of rocks, green sticks and bark.
- Bake breads in ashes by making balls of dough, covering them with flour and putting them in ashes.
- Boil water in a concave rock or bark dish by placing clean, heated rocks into water until it boils. Place food in container of boiling water, cover with large leaves, let set for an hour or until food is well done.
- Dig a hole into the side of a bank and start a fire in it to preheat the opening and surrounding soil. Place food back behind the fire as in an oven.
- Steam food with heated stones (clambake style). Dig a hole and line it with rocks. Build a fire in the hole and allow it to burn down into ashes and hot coals. Remove ashes and coals and place food directly on and between stones. (This method works extremely well for shellfish and similar foods.) Cover food with whole plant leaves, green grass or seaweed, and cover leaves with sand or soft soil. Poke small holes in the dirt and pour water in to create steam. Plug the holes to prevent steam from escaping. Cooking time depends on quantity of food being cooked and relative size of items. An hour and a half should be sufficient for most small to medium items.
- Use above method without putting holes in soil.
- Boil water in a paper cup or container over a bed of coals.

Preserving Excess Food

Carry extra food if moving from one place to another. Soft berries or fruit can be wrapped in leaves or sphagnum moss to keep them intact. Shellfish, crabs, and shrimp can be carried in wet seaweed. Fish should be cleaned immediately, washed well, and carried on a line over a pole.

Excess fish can be split (cut off head and remove backbone), spread apart and cut thin, then dried over smoke fires; spread on hot rocks; or, hung to dry from branches in the sun. Sea water can be splashed on to salt the outside. Seafood cannot be kept unless well dried and salted.

Meat can be preserved as dried "beef" or jerky (jerked meat) either over a slow fire or in the hot sun. All drying meat should be hung high to avoid animals, and covered to prevent blow-fly infestation. If mold forms on the outside, it can be brushed or washed off before eating. It may be necessary to re-dry smoked or air-dried meat to prevent molding in damp weather.

To preserve cooked animal food, it must be recooked each day, especially in warm weather. In deserts, fresh meat can be wiped dry with a cloth, cut in arm-size strips, and buried raw and uncovered, without salt in dry sand six to eight inches deep. Meat prepared this way will keep for at least three years. It resembles dried beef. To utilize the meat, it can be soaked in water to soften it and remove sand.

Drying Plant Food

Plant food can be dried by wind, air, sun, or fire with or without smoke. The main object is to remove water. Vegetables, bananas, tubers, leaves, berries—in fact most wild fruits—can be dried by cutting them into thin slices and placing them in the sun. Fire can be used if necessary.

Notes:

CHAPTER VII
Some Considerations for Emergency Environments

Desert Environments

During the extreme heat of the day find shelter in the shade. At night, seek protection from the cold. Deserts also have extremes of cold in the day during the winter. Water is a prominent concern. Water carried in the aircraft will often be the primary source. Solar and vegetation stills provide additional sources.

Insects and animals to be aware of are primarily crawling type insects and snakes. The heat will generally keep these critters inactive during the day. Evening and early morning are the times of their greatest activity. They are of no concern in the cold.

Clothing should be light in color to reflect the sun. Wear it loose fitting and in sufficient layers to combat cold at night. In winter wear dark-colored wool, and dress in layers.

Activities should never be carried on during mid-day unless during the winter. Ration acitivities not water. Little or no firewood or burning material is available on the desert. Use signals dark in color to contrast with the light environment.

Winter and Artic Type Environments

Shelter considerations are for protection from the cold and insulation of the body. Snow is probably the most efficient insulation material available because of its trapped dead air spaces. Water is of critical concern because hypohydration is much more subtle than in the desert. Primary source must be in the aircraft, or procured through some type of heat source. If the weather is clear, snow can be melted on black plastic. Animal life is limited and confined to relatively small areas.

Clothing should be dark in color to absorb heat, wool in fabric or have adequate insulation. Synthetic fibers are best in a variety of environments. Wear in layers and add sufficient layers to protect the body through the night. Improvise clothing from the cabin interior. A hat is a must — preferably a stocking cap.

Firewood and burning material is prevalent in mountainous areas but scarce in many lowland areas during winter. Signals must be dark in color to contrast with the light surroundings.

Salt and Fresh Water Environments

Flotation is the number one concern, either personal or as a group (PFD versus raft). In colder climates, insulation and adequate body protection are essential in flotation.

Shelter considerations immediately are for protection from cold water and immersion hypothermia. The salt water itself is a secondary concern. It will cause saltwater sores in a relatively short time. Guard against the sun, both reflected from the water and direct, for sunburn as well as injury to the eyes.

Signals should be dyes, streamers, bright colors, mirrors or pyrotechniques that contrast with the dark color of the water. The only heat source may be insulation, or a small catalytic heater such as a hand warmer.

Water is of critical concern in the salt water environment. Desalting kits and solar stills are a must on equipment lists. The initial source will have to be what is in the aircraft. Plastic tarps or paulins can be used to catch rainwater. Motion sickness pills are very important. Vomiting increases the hypohydration process drastically.

Clothing should be insulative in quality (even when wet), nonabrasive, and able to be sealed relatively well around the sleeves, collar, waist, and legs. Dress in layers for flotation and insulation. This is to prevent free flow of water to the body. Covering for the head is a must. In colder climates, clothing should be dark in color to absorb heat from sun, light in color in warmer areas to reflect sun. Sunburned skin, parch and chapped lips, face and extremities are usually the rule. Some type of skin ointment, chapstick or salve is necessary.

Animal life considerations in fresh water are not of primary concern since there are very few fresh water bodies of water that are of significant size to warrant extended time periods in the water. In salt water, consideration should be given to sharks and other similar marine life. Very little is known about actually predicting what these creatures will, or will not do, when exposed to humans in their environment. In general, sharks seem to feed most actively during the night and particularly at dusk and dawn. After dark they show an increased tendency to move toward the surface and into shallow shore waters. The human body's strange appearance probably alarms a shark. There is evidence that a group of clothed people in the water are safer than individuals. A shark is attracted by weak fluttery movements and may be repelled by strong regular movements and certain loud strange noises. Scent items such as garbage, body wastes and blood stimulate them to explore in the area for food.

Sea Coastal Environments

A unique consideration of the coastal environment is the procurement of water. The primary source (outside of that carried in the aircraft) will be through a combination of desalting kits, solar stills, and beach wells. The latter will produce the most. A coastal seashore is the most conductive to survival of any environment.

Extreme Northern Latitude Environments

Shelter considerations during winter are for extremes of cold. In summer, it is imperative that protection from insects be provided. It has been estimated in some areas of northern Canada that mosquito and biting insect populations' total mass outweighs the total mass weight of all large game animals. They can drive a person completely crazy. Keep the body covered with clothing or improvise mosquito netting. Consider the use of smoke to rid insects.

Water is generally available during the summer months but must be procured with some kind of heat source in the winter. The primary source during winter must be through melting or what is available in the aircraft. Keeping water unfrozen will be a problem. Firewood and burning material is in many cases scant, and very small in size. In general, there should be ample supply for fueling a fire.

Clothing should be dark in color and thick enough to withstand insects. Wear in layers. Foot travel anywhere in the vicinity of the aircraft may be restricted due to the muskeg during the summer. Floating clumps of vegetation in a bog-like area may be the only surface.

Important Survival Factors

Location

Always try to improve present location rather than moving to a new one, unless some new event has made the present location dangerous. Even then, move only after surveying a new site. In hot weather, look for shade; in cold weather, find protection—travel only as a last resort.

Weather

In bad weather, find shelter immediately as animals do; watch for weather indicators, prepare for the worst, wait it out. Never travel unnecessarily in bad weather.

Darkness

Prepare for darkness while it is still light. Darkness masks landmarks, even in familiar territory. Curtail night travel except for life or death situations. Temperature drops drastically at night in desert and cold environments. In hot climates, hold off on heavy work until the cool of the night. In cold weather, restrict or limit shelter area to a minimum for heat conservation.

Wind

Wind is the greatest source of body heat loss through convection. In cold weather, protect against wind at all costs. If windproof clothing is not available, improvise with plastic bags or anything nonporous. In desert or hot climates, wind tends to increase perspiration evaporation, hastening dehydration.

Heat

Hot weather and hard work compound body heat problems. A six degree gain in body core temperature curtails rational thinking. Stay quiet during the hottest part of the day. Desert temperatures are hottest at the surface. Temperatures two feet above or below the surface can be 30 to 40 degrees cooler. Desert shelters should provide shade and have floors at least two feet above the ground surface. Digging to get two feet below the surface might use more energy than it saves. In hot weather, ration sweat, not water. Take special care in hot climates to maintain body chemical balance.

Sun

Avoid overexposure to sun and try to find shade. On snow or water, protect face, eyes, and body. In cold climates, sun provides warmth for the body and melts snow for drinking water. The sun also aids in direction finding.

Cold

The body loses heat through five main methods: conduction, convection, radiation, respiration, and evaporation. A six degree drop in body temperature leaves us unable to help ourselves. In cold climates, a shelter should be as small as possible to conserve body heat rather than heating useless space. Since basic metabolic rate drops to 80 percent while sleeping, never sleep in cold weather without adequate shelter.

Rain

Most clothing loses almost 90 percent of its insulation value when wet. In a damp situation, seek shelter immediately and put on rain gear or protective clothing before regular clothing becomes wet. Wear wool garments outdoors. Wet cotton extracts heat 240 times faster than dry clothing.

Lack of Visibility

Stay put and restrict movements to a small area to prevent injury or becoming lost.

Physical Dangers

Be keenly aware of possible hazards from above, below, and the sides. It makes sense to expect the worst and hope for the best.

All of these hazards attack the body in numerous ways. Some work alone, some combine to worsen the situation. One or more unchecked or ignored could make the difference between life and death. Read body indicators to continue at the best mental and physical efficiency under all conditions.

CHAPTER VIII
Emergency Preparedness Kits

Basic to anyone's capability to cope with an emergency is a versatile collection of resource material which provides the essentials of life. It is imperative this material provide means to notify others that someone is in distress. Popular terms for this collection of resource material might include *survival kit, emergency kit, life support kit, or storm kit*. Essentially they provide an individual with resources to cope with the situation.

What should a kit contain? There is no universal kit that will suffice for all types of situations in all locations. Every kit manufactured was engineered to aid an individual in a particular environment. Each location has its specific obstacles. These regionally specific problems provide the basis for the large diversity in emergency preparedness kits. The best kit is one that an individual makes for his own use. This provides the familiarity for proper utilization of a kit. It also provides items an individual is personally comfortable with in a variety of situations. Kits of this nature will vary with every individual. There are several basic considerations.

Do not be confused with the term "kit". Materials mentioned here do not have to be contained in what would ordinarily be termed a kit. It refers to the centralizing of materials into one location.

Search and rescue statistics indicate long-term survival situations are rare. Case histories show most people are found in less than 72 hours (the majority are found within 24 hours). An emergency preparedness kit should provide necessities of life for 72 hours or less. Each geographic location presents problems, and an emergency kit must suit personal needs.

Basic guidelines for most environments include the following:
- Instant body shelter — Plastic tube, bags, or sheet to retain the body's vital radiated layer of warmth by reducing or eliminating convection currents. Keeping clothing dry reduces the possibility of hypothermia because one of the prime prerequisites for this deadly killer is wet clothing.
- Fire making capability — Matches, metal match and cotton, or flint and steel and some other type of tinder, and a candle. The candle is an extension of the match to dry tinder and help start the fire. Fire provides supplement to body heat, warms drink and food, and provides a tremendous morale booster. It can also be used to signal and to dry wet clothing.

- Inducement to drink hot liquids — The container in which the kit is packed can be used to heat water for warming the inner body core. Hypothermia is the number one outdoor killer. Drinking hot liquids maintains body core temperature and helps eliminate hypothermia. Tea, chocolate, coffee, and soup make hot liquids more attractive and provide a source of instant energy.
- Signal capability — Mirror, whistle, or bright-colored cloth. Anything to alert others of distress. Signals make any group more visible.

Beyond the above items, anything additional would be personal preference, based on problems any specific area might present.

Think! Your brain is your best survival tool!

The first few hours of any wilderness survival emergency will be the most important. Decisions made during this period usually determine life or death.

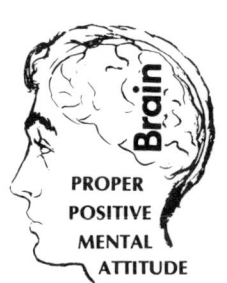

PROPER
POSITIVE
MENTAL
ATTITUDE

Aircraft Emergency Kit
Body Shelter

- ☐ Plastic tube tent.
- ☒ Large garbage bags (7 bushel).
- ☐ Plastic tarp (9' x 12').
- ☐ Space blankets.
- ☐ Mosquito net.
- ☐ Insect repellent.

Life Support Tools

- ☒ Hack saw—single handle, wood blade and metal blade.
- ☒ Plier—vise grip.
- ☐ Plier—slip joint.
- ☐ Screwdriver set (multiple).
- ☐ Cable (wire) saw.
- ☒ Knife.
- ☐ Compass.
- ☒ 50' nylon line.
- ☐ 24' small wire (brass).
- ☐ Fish line and hooks, lures.
- ☐ Desert/vegetation still material: clear poly bag, surgical tubing.
- ☐ Penlight, with extra batteries and bulb.

Vise Grip

First Aid Kit—Personal

- ☐ Sealable plastic container.
- ☐ Two (2) compress bandages.
- ☐ One (1) triangle bandage.
- ☐ Small roll 2" tape.
- ☐ Six (6) 3 X 3 gauze pads.
- ☐ Twenty-five (25) aspirins.
- ☐ Razor blade or scissors.
- ☒ Hotel size soap.
- ☐ Kleenex—purse size, or toilet paper.
- ☐ Six (6) safety pins.
- ☐ One (1) small tube of Unguentine or Foile.
- ☐ Ten (10) band-aids.
- ☐ Chapstick.
- ☐ Six (6) butterfly bandages.

Fire Making Capability

- ☒ Matches—strike anywhere—waterproofed.
- ☐ Candle.
- ☐ Flint and steel.
- ☐ Metal match.
- ☐ Magnesium fire starter.
- ☐ Dry tinder (cotton balls).

Signalling Capability

- ☒ Mirror.
- ☐ Signal panel—red or yellow.
- ☒ Flares.
- ☐ Canned smoke.
- ☐ Whistle.
- ☐ Strobe light (extra battery).

Food and Energy Package

- ☐ Container — any lightweight metal container with lid, suitable to heat and store water.
- ☐ One-man, 5-day rations, 2 or 3 cans Sego, Metrecal for liquid and energy.
- ☐ Thirty (30) sugar cubes, wrapped.
- ☐ Ten (10) packets salt.
- ☐ Ten (10) pilot bread or twenty-five (25) crackers.
- ☐ Five (5) sticks gum.
- ☐ Ten (10) bouillon cubes.
- ☐ Twenty (20) protein wafers (if available).
- ☐ Put each item in small plastic bag and seal.
- ☐ Three (3) tea bags.
- ☐ Twelve (12) pieces of rock candy.
- ☐ Water purification tablets (iodine, halazone).
- ☐ Put everything in small metal can (cook pot), seal with poly bag, and tape.

Rat Trap

Supply Sources for Emergency Equipment

SDU-5/E—(Light marker, strobe type for distress). Manufactured by FED of Handicapped, New York 10011. Also available through Eddie Bauer of Seattle and Bren-Tronic, Inc., East Northport, NY

Hand held smoke cartridge(30 second burning time)—For use with Pengun Launcher. Pengun Industries, Inc., Parkesburg, PA 19365.

Pengun Launcher and cartridge flares—Available through: Penguin Industries, Inc., Parkesburg, PA 19365.

MK-13 Day/Night Flare (Mark-13)—Manufactured by Kilgore Corporation, Hiway 18, Toone, Tennessee 38381. Distributed nationwide by marine equipment supply companies and emergency equipment distributors.

CYALUME Lightstick (chemical light)—Distributed nationwide by American Cyanamid Co., Organic Chemicals Division, Bound Brook, NJ 08805.

Skyblazer Aerial Flare Launcher (Survival Systems, Inc.)—Now distributed by Sigma Scientific Inc., 1830 South Baker Ave., Ontario, California 91761.

Signal Mirror (bright sport aiming system)—Sigma Scientific Inc., 1830 South Baker Ave., Ontario, California 91761.

Signal Mirror (small effective glass military type)—Revere Glass Co. Inc., Revere, MA.

Emergency Signalling Mirror (ESM/2)—Large military issue type glass mirror with center cross sighting system. Manufactured by General Electric Corporation, address unknown.

Gyrojet Flare Launcher with cartridges—MB Associates (MPA) San Ramon, CA.

Sport and Survival Provisions—Distributed by Svein Madsen LTD, Import and Wholesale Marine Hardware, P.O. Box 34123, STN. D Van., BC V6J4M1 (A product of Norway).

Seven Oceans Emergency Rations—Produced by Compactas, Bergen, Norway. Distributed by Svein Madsen LTD, Import and Wholesale Marine Hardware, P.O. Box 34123, STN. D Van., BC V6J4M1.

Emergency Fire Kit—Distributed by Redi Fire, P.O. Box 3307, San Bernardino, CA 92404.

All Purpose Emergency Overnight Stormkit—Mountain Rescue Council, Tacoma Unit, Tacoma, WA.

Signal, Survival and First Aid Kits—Available through Sigma Scientific Inc., 1830 South Baker Ave., Ontario, California 91761.

Survival and First Aid Kits—Available through Survivit Company, P.O. Box 334, Lake McQueeney, TX 78123.

Emergency Safety Kits—Rockford Safety Equipment Co., 4220 Hydraulic Road, P.O. Box 5166, Rockford, IL 61125.

Custom Signal, Medical and Survival Kits—Ed Delaney Associated, 1129 Paradise Parkway, Tacoma, Washington 98466.

Coldwater Survival Suits—Imperial Manufacturing Co. P.O. Box 4119, Airport Industrial Park, Bremerton, WA 98310.

Survival Kit and Emergency Stove—"Little Demon Survival Stove Kit", Survival Education Association, 9035 Golden Given Road, Tacoma, WA 98445.

"Little Demon" Special Forces Combat Stove—With Heximine fuel tablets, Survival Education Association, 9035 Golden Given Road, Tacoma, WA 98445.

Global Survival Kit with Stove "Little Demon"—Manufactured by Eddie Bauer of Seattle and Survival Education Association of Tacoma, WA 98445.

Sea and Water Survival Equipment—Available through Eastern Aero Marine (air, sea, rescue aids and aviation Products). P.O. Box 593513, Airport Mail Facility, Miami, FL 33159.

Sea and Water Survival Equipment—Also available through Avon Raft Distributors. Check your local area for distributors.

Cold Water Flotation and Insulated Jacket—U-Vic Jacket, University of Victoria, Victoria B.C. Canada V8W2Y2.

Flexible Steel Pocket Saw—Varco Inc., Belleville, NJ 07109.

Note: This is a listing of those survival equipment companies that we are aware of. We are not endorsing any product or company, but merely providing a list of equipment sources for your information and convenience.

CHAPTER IX
Search and Rescue

Very few aviators know and understand the search and rescue (SAR) systems that provide the response for overdue/missing civilian aircraft. For the most part, only the military gives its pilots information and orientation about SAR during flight training school. The majority of pilots who fly for fun or business are not well informed about SAR response, though the information is readily available if an effort was made by inquiring to appropriate agencies. Unfortunately most pilots will learn about aviation emergency response when it is too late: a friend is missing or overdue; an ELT activates accidently; the pilot is forced to make an emergency off-airport landing.

> *All those who fly or ride as passengers should have a basic understanding of what to expect in an emergency situation, and, how they can more effectively aid in their own rescue or properly report search or rescue incidents.*

Search and rescue in the United States is based on the humanitarian principle which compels people to render aid to those in distress. These distress situations range from the emergency transportation of a premature infant to an extended search for a missing aircraft in an area covering several thousand square miles. Each situation is unique and requires investigation to determine the last known location, the urgency of the situation, and the number of people involved. Then SAR forces must be logically selected to render aid to the distressed personnel.

Search Verses Rescue

Traditionally, both elements of search and rescue have been considered together. For example, a ship is overdue, a search is launched and the people aboard are rescued. Or an aircraft is overdue, its fuel presumed exhausted, and again a search is launched with hopes for a rescue. In both aeronautical, maritime, and inland operations, search and rescue are two parts of one story. We believe it is appropriate to consider search and rescue separately—as functions in many ways dissimilar, although related.

State-of-the-air differences account for this suggested separation. Basically rescue skills, equipment, resources, and trained personnel are available to respond to the spectrum of rescue emergencies. Rescue is commonly defined as "a known subject, in a known location, whose condition is described as time-critical due to injuries and weather". Rescue incidents require a ready response, quick action, and possible removal of the subject to definitive medical care. Generally, rescue takes only a few people having a high degree of training, specialized equipment, effective communications, and rapid, safe transportation.

The term "search" denotes that someone is lost and disoriented on the earth's surface and due to weather, terrain, physical condition, or injury and, due to a lack of experience and proper equipment, may be in life-threatening danger. A search effort comprises a multitude of investigative procedures varying from probabilities and statistics, interviewing techniques and background information collection to sophisticated electronic tracking and directional equipment. Each procedure will be discussed here in somewhat limited detail. Many individuals have complicated and added to the difficulty of an already tenuous search strategy with their wrong actions due to ignorance of search operations. The primary purpose here is to impart information which has repeatedly proven crucial in SAR missions, especially relating to the searching for and the rescuing of people missing or overdue in aircraft.

> **If you know where the aircraft is—it's a rescue. If you don't—it's a search.**

The National Search and Rescue Plan

Although there are many agencies and volunteers involved with SAR across the nation, the federal government assumes some responsibilities for overall coordination, especially the coordination of any federal or military resources which may be requested to respond by local or state agencies. The National Search and Rescue Plan identifies federal responsibilities and is the basis for the **National Search and Rescue Manual** that discusses search and rescue organization, resources, methods, and techniques. Though guidance is provided by the federal government, local and state government agencies are expected to assume the responsibilities for initial SAR response commensurate with their capabilities and within their geographic boundaries. In general, the federal role is one of coordination between local, state, and federal agencies and to create a cooperative national SAR network.

According to the National SAR Plan, all maritime or navigable water SAR is the responsibility of the U.S. Coast Guard. All inland SAR is the responsibility of the U.S. Air Force. These two military organizations will be the federal coordinating agency for federal resources responding to SAR incidents within their respective areas of responsibility.

Air Force Rescue Coordination Center

The Aerospace Rescue and Recovery Service (ARRS) operates the Air Force Rescue Coordination Center (AFRCC) which is the single federal agency responsible for coordinating search and rescue activities in the 48 contiguous states. The AFRCC's prime mission is the coordination of SAR, both for military and civilian personnel.

The AFRCC is centrally located at Scott Air Force Base, Illinois, 20 miles east of St. Louis, Missouri. It is operated 24 hours a day by personnel trained and experienced in SAR operations. The Center is equipped with excellent telephone, teletype, and hot line communications. A resource file lists all federal, state, local, and volunteer organizations capable of conducting or assisting SAR operations. However, the AFRCC is not authorized to commit federal funds to hire SAR resources. In addition, a listing of Mexican and Canadian SAR coordinating agencies is available. The Center is administratively divided into three sections; an operations section to prosecute individual SAR missions, a directorship to provide overall management and formulate SAR plans, agreements, and policy, and a reports section to maintain data and records.

Federal Aviation Administration

The Federal Aviation Administration (FAA) through its Air Route Traffic Control Centers and Flight Service Stations, monitor and flight-follow aircraft filing flight plans in the Inland Region. In some cases, individual citizens contact an FAA facility when they have knowledge of a probably SAR situation involving aircraft. Therefore, FAA is usually the first agency to alert the AFRCC on an emergency or overdue aircraft. The AFRCC is tied directly into FAA's teletype network, and FAA facilities use the teletype to initially alert the AFRCC.

Once the AFRCC is alerted, the FAA and AFRCC work together in trying to determine the urgency of the situation. Initially, a review of all radio communications is accomplished with the objective of ascertaining as closely as possible the last location of the distressed aircraft. Concurrently, other FAA facilities begin a check of all possible recovery airports for the missing aircraft. The AFRCC, in the meantime, contacts relatives and friends of the pilot or passengers aboard the missing aircraft, with the hope of establishing the whereabouts of the aircraft, or to gather information on the personnel aboard. The AFRCC attempts to get a description of the aircraft, its capabilities and nuances, and also data on emergency equipment aboard and to determine the pilot's intentions. Through experience, the FAA and AFRCC have learned that the majority of alerts for missing aircraft are generated by failure of the pilot to either close the flight plan or failure to specifically inform some person or agency of his/her intentions. For this reason, only a small percentage of alerts issued by FAA result in an actual airborne search for a missing aircraft.

With the recent enactment of federal law requiring most aircraft to be equipped with an ELT, the AFRCC works very closely with FAA to readily locate the source of ELT signals. All ELT signals reported to FAA facilities are immediately forwarded to the AFRCC and jointly investigated by the AFRCC and FAA as probable distress signals. However, due to ELT equipment malfunctions and presumably pilot unfamiliarity with the equipment, the majority of signals reported are not related to a distress situation. Furthermore, preliminary evidence of crashes of aircraft equipped with ELTs seems to indicate the equipment lacks the survivability necessary to perform as a completely reliable crash location device.

State Agencies

A number of states, especially in the Pacific Northwest, have established by law a responsible state agency for directing and coordinating air SAR activities. These State Departments or Divisions of Aeronautics develop and maintain their aviation search and rescue response programs with cooperation and support from local and federal agencies. It has been our experience that this system works far superior when compared to those other areas of the country that rely on the federal government to initiate and carry out aircraft SAR activities.

Civil Air Patrol

Almost consistently throughout the United States, the Civil Air Patrol provides the bulk of the response to downed or missing aircraft situations. They are a private, non-profit corporation of volunteers devoted to assisting aviation by providing emergency response and aviation safety education. Upon request the Civil Air Patrol will provide to the appropriate authority in charge of the air search or rescue effort: mission coordinators; aircraft, pilots, and observers; ground search teams; base camp support; communications networks. When officially tasked and involved in a search or rescue mission, they are reimbursed by the U.S. Air Force for communication expenses, and fuel and oil expenses incurred by aircraft or ground vehicles. In addition, because they are an official auxiliary of the Air Force, all Civil Air Patrol members are covered by the Federal Workman's Compensation Act in the event of an injury. Current statistics show that Civil Air Patrol members respond to three-fourths of all air SAR missions.

The Search Effort
How Does the Initial Report Come In?

There are a number of different ways in which a potential missing or overdue aircraft mission is initiated and handled. FAA through its *air route traffic control centers and flight service stations* monitor and flight-follow thousands of aircraft that file flight plans. In those cases where flight plans are not filed, an ELT transmission, a radio distress call, or a phone call from a concerned friend or relative will generate some type of action and follow-up. If a pilot files a flight plan establishing the itinerary, and subsequently fails to appear at the airport and time specified, the air SAR system for that area will be alerted and activated.

> *The filing of a flight plan is a pilot's best rescue insurance policy.*

Phases of the Initial Search

When an aircraft on a VFR flight plan fails to arrive at its destination, the flight service station servicing the destination airport issues an information request (INREQ) to all the flight service stations along the intended route of flight and to AFRCC. They request any information concerning the overdue aircraft. This is the first of three phases in an incident; the uncertainty phase. The second phase is the alert and distress is the third phase.

During the uncertainty phase (INREQ), a preliminary communications search (PRECOMM) is conducted by FAA, the AFRCC, and state agencies. The INREQ is issued one hour after the elapsed ETA on a VFR flight and 30 minutes on an IFR flight. If the information request produces negative results, the incident then progresses from the uncertainty phase to the alert notice (ALNOT) phase. The ALNOT is issued by the responsible FAA flight service station and is issued to cover an area normally 50 miles either side of the intended route of flight from the last known position to the destination. It can also be extended to cover the maximum fuel range of the aircraft. There are no set rules as to when an alert notice will be issued. Generally, an ALNOT will be issued at the completion of the preliminary communications search, or at the time of aircraft fuel exhausion, or any time there is serious doubt as to the safety of the aircraft or its occupants.

After the ALNOT is issued, an extended communications search (EXCOM), is conducted by the AFRCC and other agencies. The flight service station processing the alert notice is responsible not only for a communications search, but also to insure the airport personnel in their areas physically check their ramps and look for the overdue aircraft. During the EXCOM, the AFRCC and state aeronautics agencies will notify law enforcement agencies in the areas of highest probability that the aircraft is overdue, and request any leads concerning the aircraft be forwarded to the AFRCC or state agency in charge.

During the EXCOM phase, the particular agencies are extremely busy. They contact parents, business associates, friends, relatives, and owners of the aircraft gathering all the possible information on the person or persons missing. Habits are questioned and experience level checked. Weather services, both civilian and military, are queried for all the weather information along the intended route of flight. Weather is always a significant factor. The mechanical soundness of the aircraft is obtained along with questioning the owner as to radios, navigation aids, survival equipment (to include ELT) status. "Line boys" are questioned and tower operators are contacted. Did the pilot have an alternate plan to going from "A" to "B"? How about a possible point "C"? How long was the pilot going to be gone, who or what was the pilot going to see at the destination?

Also during the EXCOM, the AFRCC and state agencies monitor the actions of the FAA and keeps them advised of any findings. In short—they are gathering all the information on the pilot and aircraft, hoping to make a quick "find." They normally locate most of the ALNOT objectives in this manner—it's usually the case of not closing out a flight plan properly, or landing at an airport other than the filed destination.

If the results of the INREC and ALNOT are negative, a full-scale search (the distress phase), both by aircraft and ground teams, is initiated. More often than not, the unfortunate circumstances of this scenario is that it is carried out late in the day. The complications of bad weather, darkness, and lack of information and clues mount up quickly.

Important Note: Unless there is a functioning ELT alerting ground rescue or if weather and dark permits air rescue to the distress location, air search efforts will not begin until the weather and daylight permit safe conditions for search aircraft. Even with a functioning ELT, weather and terrain may slow down response time. With these facts known, it is imperative that all pilots and passengers receive emergency preparedness training and that all aircraft carry emergency preparedness (survival) equipment.

The Search Mission: Solving the Classic Mystery

During the first few hours of the search, a multitude of facts and information is needed to piece the puzzle together. Was there an Inflight Emergency (IFE) reported? If so, where was the first known location? Have all air carriers and flight service centers been asked for ELT reports? If there is an approximate location and time for the last communication with the missing aircraft, then is there a record of when the aircraft went off of radar on the Interim Track Analysis Program (ITAP)? If the emergency could have been weather related, are there satellite pictures available showing weather disturbances and fronts at the time and location of the probable emergency? All of these questions and more must be answered as quickly as possible.

Mission coordinators for air search begin with generalities. If the flow of information and clues are complete and accurate, the length of time to locate the missing aircraft will be short. If not, the search effort can be long.

Planning a Search

The planning of a search involves estimating the most probable position of a distress incident or of its survivors; determining a search area large enough to assure that the survivors are somewhere within the area, choosing the equipment to be used in the search, and selecting search patterns to be used to cover the area. Some SAR operations present little or no difficulty as to the search phase. The incident may have been witnessed, or its location may otherwise be known accurately enough to preclude the need for a search or to present any difficulty in determining, within narrow limits, the area to be searched. In other cases, however, the determination of a search area assumes a very different character. Lack of sufficient information as to the position of the incident or survivors will make it difficult to determine the areas which should be searched first.

The Possibility Area

The area of maximum possibility is roughly a circle with its center at the last known position of the aircraft and a radius equal to the endurance of the aircraft at the time of its last position, expressed in terms of distance as affected by wind velocity. It involves the assumption that the aircraft may have flown from its last known position until fuel exhausted, on any course, even at right angles to, or the reciprocal of, the intended flight plan.

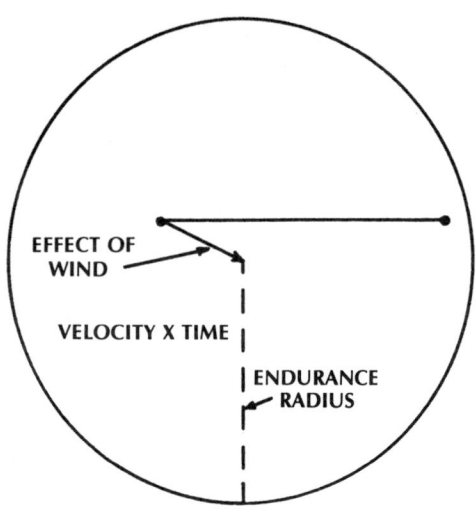

Systematic search of such a large area would not be conductive to efficiency since it would not permit the mission coordinator to concentrate the initial search effort in the area or areas that the missing aircraft would most likely be. Determination of the maximum possibility area will, however, enable the mission coordinator to screen hearing and sighting reports coming in as leads.

The Probability Area

Determination of the probable area, i.e., the area in which the aircraft is most likely to be, is based on the degree of accuracy that can be attributed to the aircraft's last known position. The position of a distress incident can be determined within fairly narrow limits when the following information is received: a location where the aircraft disappeared off radar; a bearing or fix provided by a ground station or emergency radio aid; a dead reckoning (DR) position based on time of last known position; or reports of sightings, ground or air. When this information is not available, the probable position may be narrowed considerably if information or one or more of the following items can be obtained and assessed: a flight plan or route of object; complete information on weather along intended track or route; proximity of airdrome along the track; aircraft performance; the pilot's record, habits, experience; the radar coverage along intended track (as a limiting factor); the nature of terrain along intended track; or position and ground reports (leads).

When information vital to determining the most probable position of the incident or survivors is not available, search planning becomes a difficult matter. It may be impractical and time consuming to search the entire area in which the incident may have occurred or in which survivors might be located; the mission coordinator must reconstruct the incident with whatever information is available. The search plan in such cases, is usually based on the presumption that the aircraft met with an accident, became lost, or was forced down near its intended course. The initial phase of search is confined primarily to the intended course and its surrounding area. If no results are obtained, the mission coordinator must either extend the initial search area or determine other areas based on hypothesis.

Unless there is information to the contrary, it is generally assumed that the most likely area within which a missing aircraft will be found is along the intended track from the last known position to the intended destination or within a reasonable distance either side of a track. However, the probable error in DR estimate is assumed to be 10 percent of the distance from the last known position for aircraft. In this sense when talking about last known position, we refer to an enroute fix or the airport of departure, whichever the aircraft was last known to depart.

Probable Area—Approximate Position Known

When the approximate position of a distress is known, or can be fixed, or can be estimated with reasonable certainty, the radius of the most probable area is relatively small. However, if the information received is not complete and leaves room for uncertainty, the radius must be increased sufficiently to insure the inclusion of the

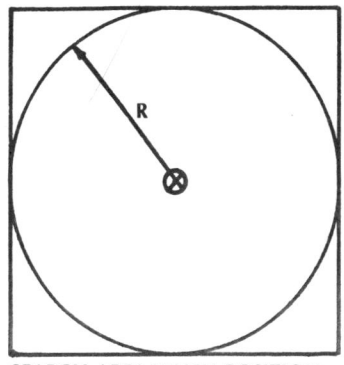

SEARCH AREA WHEN POSITION OF INCIDENT KNOWN

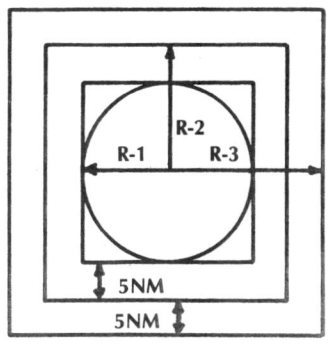

EXPANDED SEARCH AREAS

search target in the area to be searched. The circle derived should then be squared off to facilitate the establishment of search patterns.

If the search of the area proves unsuccessful, the probability area will have to be expanded gradually with radii that have been increased successively by, say five nautical miles (NM). The search is then resumed and, if this appears advisable, also repeated in the area already covered. Thus, the repeated search of the area in which the target is most likely to be, i.e., the central area, will increase the chances of its detection. It is important to note that insufficient extension of the search area may have been the reason why in some searches survivors were not found.

Time of Incident is Known

If sometime after reporting its position (last known position) an aircraft enroute between two reporting points signifies that it is going down without indicating where, the approximate position of the incident may be calculated from data extracted from the flight plan. The probability area is determined as follows: a circle with a radius of 10 NM is drawn around the last known position and another circle with a radius of 10 NM or 10 percent of the track distance in NM, whichever is greater, is drawn around the next reporting point or the final destination. Next, straight lines tangential to the circles are drawn. The area of highest probability will be somewhere in the area defined by the first circle and the tangential lines. The area of *first priority* for search will be centered on the probable location of the aircraft at the time it signified it was going down. The area of second priority then will be determined by assessing information available.

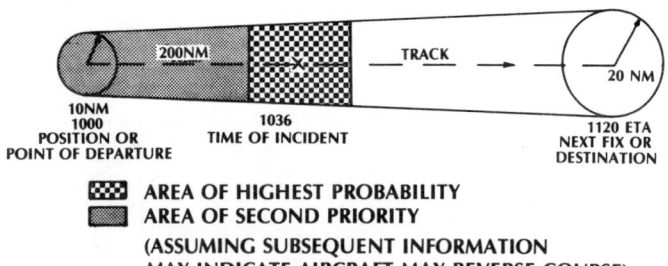

No Communication or Last Position Unknown

When an aircraft disappears enroute the first presumption normally acted upon is that it has met with an accident on or near track. Search operations in these cases are normally confined, in the early stages at least, to the immediate vicinity of the track from the last known position onwards.

Probability Area Number One

A circle with a radius of 10 NM is drawn around the last known position, i.e., a fix enroute or the airport of departure. Additional circles with a radius equal to 10 percent of the *total* track distance in NM from the last known position to each turning point and to the destination, i.e., airport of next intended landing, or circle of 10 NM whichever is the greater, are drawn around each turning point and at the destination. These circles with lines drawn tangential to them form probability area number one. The whole or any part of this may be given first priority for search dependent upon information available.

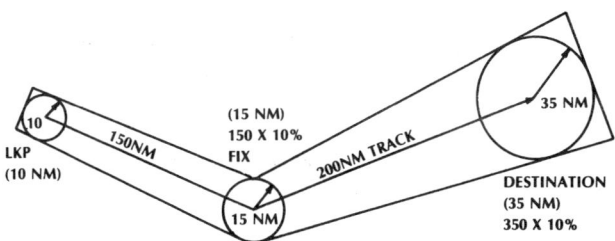

Probability Area Number Two

Plotted same as last example with radius of circle increased to 15 NM at LKP and 15 percent of track distance. Area two is the area then between boundary of area one and the new boundary (shaded area).

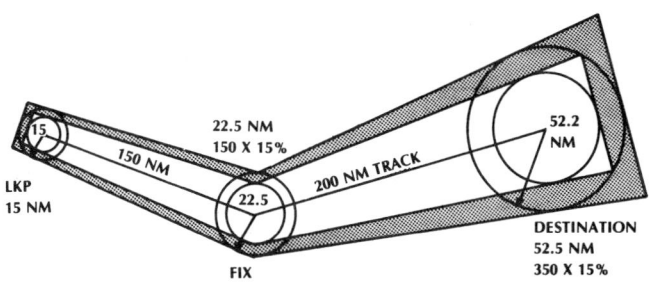

Probability Area Number Three

Plotted same as previous examples except circles of 20 NM and 20 percent track distance are used. The area between the boundary of area two and the new lines form SAR area number three.

What if The Aircraft is Not Found?

Probably the toughest question that must be answered by a search coordinator is "How long do we continue the search when nothing positive turns up?" In general, most air searches will be conducted for approximately seven good flying days. This number may vary by a day one way or another, but in most cases, one full week will be the limit. You may ask, how can anyone arbitrarily cut off a search after such a period of time? The answer must reflect the limit to the dollars, manpower, and energy that can be spent in supporting such a mission. No one likes to call off a search when there is one shred of evidence to substantiate the possibility that there may be injured survivors somewhere waiting for rescue. But, search costs money, and there is a limit to spending which must be addressed. In most cases, a combination of factors is considered. These factors are previous statistics, evaluation of search effectiveness, fatigue of search pilots and observors, terrain, weather conditions, and known facts. These all contribute to the ultimate decision.

It has been our experience that in most cases family members, friends and sympathetic fellow pilots usually carry on the search for a missing pilot another unspecified number of days. For the most part, these searches are futile effort and seldom result in anything positive. However, many times, a local unit of Civil Air Patrol along with interested local pilots will conduct subsequent training missions in the area. Although nonproductive for the most part, they do occasionally find an aircraft. The most productive time of year for finding missing aircraft is in the fall when thousands take to the woods for hunting recreation.

In looking back through past records, there are generally only three reasons why a search has been continued for an extended period of time. Political pressure has an unbelievable amount of influence on the duration of search for any individual. If the missing pilot is either politically important or related to a political figure, there is generally no limit to the length of time or the resources that may be expended in the search. Unfortunately, that is the way the system works. Another reason is connected to clues and selected information that shed light on some crucial part of the search. In simpler terms, somebody remembers a statement, a fact, or additional information which can thrown the entire search in a different direction or an entirely new geographic location. Also, private pilots flying in the suspected downed aircraft area have often times reported sightings which have proved positive. The third was mentioned above in reference to training exercises in the suspected area. All of these have proved positive to some degree.

ELTS — Some Good Points and Some Bad

It has been a decade since ELTs were made mandatory on all aircraft across the country. It may be another ten years before an ELT is developed that lives up to the expectations of its invention. In the meantime, reliability of ELTs on the market seems to be much in doubt.

In the first year of mandatory use, upwards of 8000 false activations were recorded. In the succeeding years that number tapered off to around 5000. In recent years, the number seems to have dropped to around 1500. Recent low numbers could be attributed to the FAA directives which allowed removal of the units to replace the lithium batteries. Problems range from false triggering, and not functioning on impact to self-destructing and causing extensive damage to the aircraft. It is unfortunate that the greatest percentage of ELT activations are false and the majority of pilots do not really take the time to find out what the problem is.

However, there are several hundred people in the United States today that owe their life to an ELT. Current statistics show that the average time to locate a crash sight where an ELT functions is 22 hours. By contrast, when the ELT does not function, the crash site will take approximately four days and 18 hours to find. The odds of finding survivors in that extended time drop dramatically. In spite of all the headaches, false activations and shortcomings, the ELT still saves lives and represents the biggest single factor that can lead searchers effectively to a crash site. The false activations must be lived with while working to solve the problem. There are certain precautions that can be taken which will better insure proper activation of the units.

> **In spite of its high false alarm rate, the ELT is still the most efficient SAR tool when it functions as designed. In 1978, it aided in the location of 38 percent of all civil aircraft in distress and 50 percent of all survivors.**

What is Your Choice?

It is the opinion that something is better than nothing. Although the alkaline power sources have their problems they do perform if regularly checked and changed. In the cold, lithium batteries seem to still outperform anything to date. Just remember that in an emergency, it is imperative to keep alkaline units warm. (Keep it next to your body if possible!) Perhaps some of the companies now doing research will soon come up with a unit that will solve the problem. With all of our space technology, it is sad that no one has come up with an acceptable prototype to date.

False Activations and the Triggering System

The most common form of a triggering device in today's ELT is a G-switch. As would be expected, if the switch is designed to activate with little force, there will be a great many false activations. If it is designed to trigger with high G-forces, it may not activate in some types of crashes. To date, the entire problem has not been solved. This is why it so important to post flight check your ELT. For that matter, any excessive turbulence or jolting could throw the switch in some units. Some pilots do not make a smooth landing every time.

Another problem develops with the use of portable ELTs. The pilot puts the unit in the flight bag, walks in the house and throws the bag on the floor of the closet. The transmitter goes off and now there is a very expensive search established to find and shut off that transmitter. This situation has happened many times. We may be able to design an ELT to withstand a crash but overcoming pilot ELT carelessness could be another matter.

Another Drawback is the Antenna

There are many cases on record where the ELT survived the crash, the batteries were still functional and the unit did not work because the antenna or some part of that system was severed. Many ELTs are designed to mount in the interior portion of the rear fuselage with the antenna rigidly mounted on the outside of the aircraft. The connecting link is a piece of coaxial cable. The problems that result from this design are many.

The average ELT weighs approximately 2.2 pounds and during a 9-G deceleration crash force it may develop many thousands of pounds of instantanous forward motion. Many ELTs are mounted on the skin of an aircraft with tiny sheet metal screws. They simply are not strong enough to withstand the thousands of pounds of sheer force developed by crash deceleration and impact. The expected result is a detached ELT which quickly severs the coaxial cable in its flight and even though the unit may be operating, it is not broadcasting because it has no antenna. Have you checked the mounting and location of your ELT lately? Is it mounted on or behind a bulkhead?

The Power Source for Your ELT

As would be expected with the use of any battery operated product, there have been some reliability and maintenance problems. A major source of the battery problems associated with the ELT seem to have developed with the widespread use of the lithium battery. Although they have their attributes, lithium models on the market today leave much to be desired.

Besides just not functioning, faulty batteries have virtually gassed occupants of some aircraft, damaged interiors and control cables, and even caused explosions which have ripped holes in bulkheads and the aircraft skin. In addition, it seems a great many of the false activations can be traced to some type of short circuit caused by corrosion from the battery.

As a result of the torrent of problems with the batteries, the FAA ordered all lithium batteries out of service within 30 days and pilots with those units could be exempt from carrying an operable ELT for up to 180 days. However, there is another technical standard order which has been released on the batteries and the hope is that this will be in effect before the ban is lifted. There are those who could think of ways to use this to remove the ELT from their aircraft indefinitely. The problem of air search is difficult enough without realizing that there could be thousands of aircraft flying around with no ELT at all. Reliable or not, those units can and have saved lives.

The other part of this design concept is the rigidly mounted external portion of the antenna. Aircraft that slide through trees, or invert after impact often break this antenna off or smother it against the ground. Advocates of having both top and bottom antennas on the aircraft have been soundly reprimanded for wanting to poke yet more holes in the skin of the airplane. Other problems with interference and the multitude of antennas already required put ELT whip location very low on the antenna equation.

There are those who advocate another innovation which could be the new third generation ELT. Since ELTs have a tendency to become dislocated and fly out of the aircraft, why not design them to function that way. The idea has a great deal of merit! There are enough statistics and case histories to tell us what they do; now all that remains is to develop the trigger, antenna, and transmitter that will function after they leave the aircraft.

If the antenna breaks, improvise one.

A very important point worth remembering is that a ball point pen cartridge, a small piece of wire, or almost any small metal point placed against the coaxial cable outlet will make an ELT functional. If you can locate the unit, improvising an antenna is extremely simple.

From a SAR viewpoint, there is also another point worth remembering. If the area around a crash sight is very steep and rocky or there are large rock formations in the area, a phenomenon called spurous radiation may result in the ELT broadcast signals. Very simply put, it is nothing more than signal bounce off of the surrounding terrain. This gives numerous as well as false locations. The only solution seems to be relocating the ELT to higher uninhibited ground or trying to directionalize the antenna.

It All Boils Down to the Pilot!

There are many pros and cons to the emergency beacon program, but no matter what the problems or solutions are that solve those problems, the ultimate factor that will change the entire picture is the pilot. If the ELT program is ever to become a complete success (and it's not going to go away!) then every pilot needs to take the program seriously. Preflight checks; postflight shut down procedure with a check of 121.5; inflight frequent check of the emergency frequency with accurate and timely reports to authorities of any activities. In short, the system will only be as good as the efforts that pilots put into it. The ELT and the system that makes it work is similar to survival training and your kit. You never need it until you need it, and then it better work because it is all you have got.

1978 Actual Distress/General Aviation Search Missions		
	ELT Functioned	ELT Did Not Function
Average Elapsed Time from Last Known Position to Location	23.0 Hours	4 Days, 17 Hours, 51 Min
Survival Rate	33%	23%
Average Total Hours Flown to Locate Objective	19.0 Hours	130.3 Hours

NOTE: The average time to crash location was within the critical 24 hour period for crash survival. However, once located, additonal time is required to reach, treat, and recover crash victims. The aviation community should take appropriate actions which can further reduce the time to locate distress-activated ELTs. These include: increased monitoring and reporting of ELTs by pilots (before, during, and after each flight); increased efforts by airport managers to monitor, report, and locate ELTs (including procuring portable ELT locating devices); increased efforts by FAA personnel to solicit ELT reports and report more useful ELT data to AFRCC; expedited ELT processing and activation of SAR forces by AFRCC; state and local officials charged with SAR responsibility and the search and rescue forces dispatched to locate ELTs must respond as though lives are at stake. They often are. Search aircraft must have direction-finding equipment and search crews must be proficient in ELT location techniques. ELT location devices should have dual frequency UHF/VHF capability. Many ELTs transmit on both frequencies. It is not uncommon for the VHF channel to malfunction, resulting in a UHF-only signal.
*Information provided by the AFRCC, Scott AFB.

Flight into Foreign Countries
Canada

Coordination of SAR operations in Canada is the responsibility of the Canadian Armed Forces. Just as in the United States, a variety of additional agencies are called upon to actively participate with the overall coordination being the responsibility of the Canadian Forces Rescue Coordination Center.

In checking with Canadian authorities, the advisability of filing a flight plan in accordance with the Minister of Transportation cannot be over stressed. All information must be listed with particular emphasis given to the route of the flight and the hours of fuel carried.

Since SAR operations begin automatically when a flight-planned aircraft is reported overdue, it is vitally important that an immediate effort be made to inform a ground station of any change in your flight plan, reason for the change, and other pertinent information. If immediate contact cannot be made with a specific ground station, this information should be transmitted blind on a common frequency.

Although the vast areas of Canada are much more inaccessible, the system which provides for SAR is very comparable to that of the United States. One final comment about flying in Canada. Be sure to know the frequencies in use for the area you are flying in, and never overlook the use of your radio if forced down. At a minimum, you should broadcast for two to three minutes every hour or when you see or hear an aircraft overhead.

Also remember that there are regulations about what should be carried on board your aircraft when flying in certain regions of Canada. A good suggestion would be to check with the Minister of Transportation before planning a flight into any part of Canada.

Mexico

Most pilots are not aware that by international agreement only five U.S. Military aircraft are allowed in Mexico at one time and that special clearances are made on an individual basis by the Minister of Defense in Mexico. Civil Air Patrol aircraft are considered military aircraft as they operate under the auspices of the U.S. Air Force. When requesting clearances for search aircraft to go into Mexico, the Air Force must go through established diplomatic channels via the U.S. Embassy to the Mexican Government. It is not unusual for 24 to 48 hours to pass before clearance is granted. United States civilian pilots should realize that in practice, there is virtually no SAR organization established in Mexico. In addition, communications are extremely limited.

United States pilots flying into Mexico who must make forced landings should be prepared for an extensive delay before U.S. authorities even will be aware of the problem. From a common sense standpoint, roundtrip flight plans along with a detailed route description are essential. In many cases, the only search resources available along the border will be volunteer ground units such as the San Diego Mountain Rescue Teams (for search in the Baja). The more information you leave behind when departing into Mexico the better your chances of being found if anything goes wrong.

Conclusion

The AFRCC has compiled statistics comparing response time from the last known position to AFRCC notification of an overdue aircraft, and the time required to locate the aircraft. The following statistical sample deals only with those accidents that resulted in active searches. These figures do not include accidents in the immediate vicinity of airports that were resolved without AFRCC participation.

Response Time vs. Type of Flight Plan			
Type of Flight Plan	No. of Missions	Average Time From Last Known Position To AFRCC Notification	Average Time From Last Known Position To Crash Location
IFR Flight Plan	151	0 hrs 43 min	9 hrs 8 min
VFR Flight Plan	213	5 hrs 21 min	38 hrs 17 min
No Flight Plan	336	32 hrs 29 min	3 days 21 hrs 52 min
Survivability vs. Flight Plan			
		IFR/VFR Flight Plan	No Flight Plan
Persons Involved		682	532
Total Survivors		261	115
Suvival Rate		38%	22%
NOTE: Survivability data time span is Jan. 1, 1977 through Dec. 31, 1978 and involves actual distress missions that the AFRCC became involved in.			

NOTE: Some materials in this chapter were adapted or borrowed from:
Summary of 1978 Search and Rescue Activity, Headquarters, Aerospace Rescue and Recovery Service, Scott Air Force Base, April 4, 1979.
Search and Rescue Coordination Training Manual, 39th Aerospace Rescue and Recovery Wing, Elgin Air Force Base, 1972.

CHAPTER X

Managing Your Emergency in Hot or Cold Environments

Considerations for Managing an Emergency in Hot Environments

Phase I—Emergency Response

- [ ] Protect and Maintain Life
    - Stay away from the aircraft until engines have cooled and fuel is evaporated.
    - Stay out of direct sunlight.
    - Improvise shelter (shade).
    - Cover the body with loose, lightweight clothing, especially the neck and head.
    - Analyze immediate dangers:
        - —intensity if heat-time of day
        - —environmental hazards
        - —amount of available water
    - Stop or minimize all muscle action during intense heat periods (usually greatest in afternoons).
- [ ] First Aid, Self Aid
    - Think before you act.
    - Be careful in removing injured from the aircraft.
    - Treat injuries.
    - Watch for symptoms of heat injuries.
- [ ] Protect Equipment
    - Everything you have, including personal equipment, garbage/rubbish, stalled transportation, and natural materials are all sources of improvising life support needs.
- [ ] Conserve Resources
    - Analyze physical resources: water, shelter, clothing, food.
    - Conserve your body's water; do not sweat, and retain the sweat in light clothing. Try not to urinate. Drink water when thirsty.
- [ ] Signal Distress
    - Make sure your ELT is working.

- Use colored flags, clothing, panels, smoke, ground signals, mirrors. Do everything possible to make yourself more conspicuous and easier to be seen.
- Use the aircraft radio, if it is still operational.

Phase II—Continued Life Support
☐ Biological (Human Body Management)
- If possible, stay in the shade. Get above or below hot ground. Air temperature is 30 degrees or more cooler two feet above or below the ground.
- Do not move about in sand/dirt storms. Cover face with a cloth. Keep back to wind. Stay in shelter.
- Drink water when thirsty. Drink all you want—do not ration. Take salt, **only** if plenty of water is available. Normal diet will usually contain adequate salt requirements.
- Do not eat unless plenty of water is available. Eating will increase metabolic heat. Digestion requires water.
- Keep mouth shut, breathe through nose, minimize talking to prevent tongue swelling and minimize water loss.
- Constantly analyze what the heat is doing to your body. Do everything possible to minimize the effects.
- Improve your retention of body water. Do not sweat. Work only in early morning or at nightfall when temperatures are cooler. Let light clothing keep perspiration near the body and minimize evaporation water loss.
- Do not travel cross-country unless absolutely necessary. A gallon of water will be needed for every 10 miles of travel. If travel is necessary, walk slowly, rest often, travel when temperatures are cooler.
- Wear dark glasses or improvise eye protection to prevent sunblindness.
- Watch for symptoms of heat injury.

☐ Mental
- People who are strangers to hot, barren, uncivilized environments may quickly grow impatient waiting for help.
- Fears can overpower good judgment replacing it with a determination to escape.
- Desert rescue teams say that many fatalities are caused by people attempting to walk out.
- Most emergencies will only be of short duration. People can live through them provided they try to help themselves and do the right things first.
- Maintain a positive mental attitude and a determination to succeed.

- ☐ Emergency Preparedness Skills
  - During emergencies you will be forced to live without the direct benefit of modern technology. You will be forced by circumstances to rely upon your own knowledge, common sense, and ingenuity.
  - You can and must improvise all basic needs from resources available.
  - Review the skills outlined in this manual. Determine what is needed for your immediate comfort and survival.
- ☐ Emergency Environments
  - Again, review the chapter on Emergency Environments. Be aware of the particular hazards posed by hot environments.

Phase III—Rescue
- ☐ Analyze the approximate length of the emergency and chances for outside assistance.
- ☐ Improve signals.
- ☐ Stay with the aircraft. Make it as visible as possible.

Considerations for Managing an Emergency in Cold Environments

Phase I—Emergency Response
- ☐ Protect and Maintain Life
  - Stay away from the aircraft until engines have cooled and fuel is evaporated.
  - Put on clothing to minimize heat loss. Wear a hat. Shelter the head and neck from wind and cold.
  - Put on or improvise wind and/or rainproof clothing.
  - Get out of the elements (rain, wind, storm, etc.)
  - Make shelter. Improvise the quickest and most adequate for terrain and weather.
  - Analyze immediate dangers:
    —severity of body heat loss.
    —environmental hazards.
    —amount of your available body energy
  - Improvise heat—fires, stoves, etc.
  - Stay dry. Do not overheat your body.
- ☐ First Aid, Self Aid
  - Think before you act.
  - Be careful in removing injured from the aircraft.
  - Treat injuries.
  - Watch for symptoms of cold injuries.

- ☐ Protect Equipment
  - Everything you have, including personal equipment, garbage/rubbish, stalled transportation, and natural materials are sources of improvising life support needs.
- ☐ Conserve Resources
  - Analyze physical resources: clothing, insulation, emergency equipment. What is available for shelter, warmth, fire signals?
  - Conserve your body's heat. Add body insulation, loose fitting clothing in layers. Close all clothing openings. Use a rebreather. (Breathe through a scarf, loose weave cloth, sock, etc.)
  - Stay put in shelter until conditions improve. Conserve body energy.
- ☐ Signal Distress
  - Make sure your ELT is working.
  - Use colored flags, clothing, panels, smoke, ground signals, mirrors. Do everything possible to make yourself more conspicuous and easier to be seen.
  - Use the aircraft radio, if it is still operational.

Phase II—Continued Life Support
- ☐ Biological (Human Body Management)
  - Do not get wet. Wet clothing loses body heat 240 times faster than dry clothing.
  - Do not sweat. Sweating indicates excessive energy loss and wets clothing.
  - Improve shelter. Increase insulation. In extreme cold, avoid metal aircraft parts for shelter.
  - If available, nibble food.
  - Drink warm liquids.
  - Build heat sources.
  - Stay put. Do not fight a storm.
  - Wiggle toes and fingers. Do light isometric exercises.
  - Obtain water. Keep stored water from freezing.
  - Improve clothing.
  - Constantly watch for symptoms of cold injury.
- ☐ Mental
  - Stay calm. Worry causes anxiety.
  - Control your imagination.
  - Fears will overpower good judgment.
  - Most emergencies will only be of short duration. People can live through them provided they try to help themselves and do the right things first.

- Maintain a positive mental attitude and a determination to succeed.
☐ Emergency Preparedness Skills
  - Your aircraft can fulfill most of your basic needs if you think and improvise.
  - Gasoline and engine oil are fuels for emergency heaters and stoves.
  - You can and must improvise all basic needs from resources available. Review the skills outlined in this manual. Determine what is needed for your immediate comfort and survival.
☐ Emergency Environments
  - Again, review the chapter on Emergency Environments. Be aware of the particular hazards posed by cold environments.

Phase III—Rescue
☐ Analyze the approximate length of the emergency and your chances for outside assistance.
☐ Improve signals.
☐ Stay with the aircraft. Make it as visible as possible.

Notes

Bibliography

Adolph, E.F. and Associates. *Physiology of Man in the Desert.* New York: Hafner Publishing Company, 1969.

Aircraft Owners and Pilots Association. *AOPA Handbooks for Pilots.* Washington, D.C.: 1979.

Defense Flying. Department of National Defense, Air Transport Command. Canada: 1973.

Edholm, Dr. O.G. *Physiology of Human Survival.* Academic Press: 1965.

Emergency Landing Techniques in Small Fixed-Wing Aircraft. NTSB-AAS-72-3, National Transportation Safety Board. Washington, D.C.: 1972.

Fear, Gene. *Surviving the Unexpected Wilderness.* Tacoma, Washington: Survival Education Association, 1972.

Medical Handbook for Pilots. Superintendent of Documents, U.S. Government Printing Office. Washington, D.C.: U.S. Government Printing Office, 1974.

Mitchell, Dick. *Mountaineering First Aid.* Seattle, Washington: The Mountaineers, 1975.

National Search and Rescue Manual. U.S. Federal Forces, 1973.

Stoffel, Robert Skip. *Emergency Preparedness Today.* Olympia, Washington: Washington State Department of Energy Services, 1976.

Survival-Search and Rescue Air Force Manual 64-3. Superintendent of Documents, U.S. Government Printing Office. Washington, D.C.: U.S. Government Printing Office, 1969.

Thygerson, Alton. *Disaster Survival Handbook.* Salt Lake City: Brigham Young University Press, 1979.